REGIONAL
POLITICS

RECENT VOLUMES

REGIONAL POLITICS

◆

AMERICA IN A POST-CITY AGE

EDITED BY
H. V. SAVITCH
RONALD K. VOGEL

URBAN AFFAIRS ANNUAL REVIEWS 45

SAGE Publications
International Educational and Professional Publisher
Thousand Oaks   London   New Delhi

For information address:

 SAGE Publications, Inc.
2455 Teller Road
Thousand Oaks, California 91320
E-mail: order@sagepub.com

SAGE Publications Ltd.
6 Bonhill Street
London EC2A 4PU
United Kingdom

SAGE Publications India Pvt. Ltd.
M-32 Market
Greater Kailash I
New Delhi 110 048 India

Printed in the United States of America

ISSN 0083-4688
ISBN 0-8039-5890-0 (hardcover)
ISBN 0-8039-5891-9 (paperback)

This book is printed on acid-free paper.

Sage Production Editor: Diana E. Axelsen

96 97 98 99 10 9 8 7 6 5 4 3 2 1

Contents

Preface

For both of us, this volume has grown out of an abiding interest in two distinct though interrelated issues: first, the evolution of metropolitan regions and, second, the impact of political institutions on regional economies. This current work seeks to link both aspects through an investigation of the political economy of ten metropolitan regions. We have found these linkages to be strong and enduring, through in many ways unpredictable. What works in one region may not necessarily work in another or may result in radically different outcomes. Yet despite the diversity, we have also been able to discern common patterns, and we believe we have captured these patterns through a framework suggest in Chapter 1 and the perspectives offered in Chapter 12. We trust the reader will connect this material to the case studies in the remaining chapters.

The cases were chosen because they cover a spectrum of experience in regional institutions and political economy. The authors of these cases present a rich tapestry of metropolitan America, one that reveals the reciprocating influences of politics and economics and imparts distinct themes. Some regions are engulfed by endemic conflict. In their chapter on New York, Bruce Berg and Paul Kantor demonstrate how regional institutions attempt to manage economic tensions between states and localities. In that highly competitive region, institutions have become a product of the very conflicts they were supposed to control and, at times, have collapsed. In his chapter on Los Angeles, Alan Saltzstein shows how regional conflicts emerge out of attempts to control air pollution. He also points up the importance of transportation and the concomitant frictions over efforts to bring about clean air. In their chapter on St. Louis, Donald Phares and Claude Louishomme underscore how an intensely fragmented region attempts to cope with tax issues, competition for industry and resultant "place wars." Here, attempts to bring about institutional reform fall before the courts and a turbulent political process.

Other metropolitan regions spawn a different, more gentile theme. Jeffrey Henig, Donald Brunori and Mark Ebert show how Washington, D.C.'s region focused attention away from the volatile issues and toward technical cooperation in police services. Despite the fragility of the

District of Columbia's regional institutions, they have managed to hang on by siphoning federal support and curtailing the political agenda. Louisville and Pittsburgh deflect conflict in much the same way, though with one important difference—these regions focus on economic development. H. V. Savitch and Ronald Vogel show how Louisville's Compact brought some temporary relief to struggles over tax revenue, while Louise Jezierski turns to the role of the Allegheny Conference in promoting regional cooperation. In Louisville and Pittsburgh, powerful elites used institutions to bridge social cleavage.

Finally, the cases demonstrate that some regional institutions do work —albeit with occasional setbacks, limits and a narrow political scope. Miami, Minneapolis-St. Paul, Jacksonville, and Portland adopted different kinds of institutions to accommodate regional pressures and promote regional cooperation. Genie Stowers points up the very limited nature of Miami's metro system, which is best understood as a "regional county." Bert Swanson shows that consolidated Jacksonville cannot become a more encompassing regional institution and has failed to slow the growing disparities between the center and the suburbs. John Harrigan chronicles the story of the Minneapolis-St. Paul model ("widely praised but never copied") and now showing signs of distress. Arthur Nelson demonstrates the circumscribed role of Portland's metropolitan service district.

All told, the cases enable us to understand the dynamics of regional politics and draw conclusions about strategies for regional cooperations. We are grateful to the contributing authors for their care, skill, and diligence. Without their expertise and insight, we could not have proceeded. They should not, however, be held accountable for our interpretations of their cases.

We also thank John Mollenkopf of the City University of New York's Graduate Center, for greeting our work, encouraging its completion, and inviting us to present some of our findings to a conference sponsored by the Social Science Research Council and the Department of Housing and Urban Development. The Conference, held in Washington during December 1994, enabled us to crystallize our ideas and expose them to the probing criticism of other colleagues. Hank Savitch extends his special appreciation to Dave Rusk for his comradeship, intellectual acuity, and his stubborn search for regional solutions. Ron Vogel acknowledges the tremendous intellectual debt he owes Bert Swanson at the University of Florida and is grateful for the opportunity to turn the table on his former mentor in his new capacity as editor. Dennis Judd and Susan Clarke, series editors, were always patient. They consistently probed us to push our findings to the limit and helped put many of the pieces together.

Several Graduate Research Assistants worked steadfastly and artfully to gather data and make some sense of it. David Collins, now research Director for the Committee on Prevention, Education and Substance, and Kevin Dupont helped us immeasurably for more than two years. Suzanne Sexton and Geetha Suresh furnished us with some last-minute talent in putting the index together.

To my students in
New York, Louisville, and Paris

—HVS

To my wife, Jeanie

—RKV

1 Introduction: Regional Patterns in a Post-City Age

H. V. SAVITCH
RONALD K. VOGEL

City regions are not defined by their natural boundaries, because they are wholly the artifacts of cities at their nuclei; the boundaries move outward— or halt—only as the city energy dictates. (Jacobs, 1984, p. 45)

■ Old and New Studies of Regionalism

Thirty-five years ago, journalists, scholars, and policymakers bemoaned the multitude of local governments. Localities proliferated in every form; counties, multiple classes of cities, townships, villages, and special districts smothered the regional landscape. The prevailing opinion at that time was not that America had too much government, but that too many governments made effective governance impossible.

The scholarly critique was led by Robert Wood's *1400 Governments* (1961). Other writers followed Wood's lead and began to press for simpler, regionwide, rationalized government (Aron, 1969). The critique was given official credence by a series of government reports recommending measures to broaden the tax and service base of cities (Advisory Commission on Intergovernmental Relations [ACIR], 1967, 1969).

The solutions called for "streamlining" all those governments and setting up a few tiers of authority. The states and the federal government could establish mechanisms for coordination and respond with rational, broadly cast policies. Fewer localities could then adapt fewer policies to an array of metropolitan and local needs (Committee for Economic Development, 1966; League of Women Voters Education Fund, 1974).

In one of those rare historical moments, policy deliberation changed the institutional basis of some regions. Beginning in the late 1950s and through the next decade, a number of localities reorganized to consolidate, broaden, or strengthen the authority of metropolitan government. This occurred in Miami (1957), Nashville (1962), Minneapolis-St. Paul (1967), Jacksonville (1967), and Indianapolis (1969). Indeed, the movement toward some type of reorganization on a regional basis took hold in other nations. Toronto (1953), ahead of the pack, was soon joined by London (1964) and Paris (1967).

At the time, regional-type governments held great promise, but the regional ferment of the 1960s quieted down during subsequent decades. One major initiative (New York's Tri State Commission) was dismantled, though others were kept in place. By the 1990s, regionalism was again in the air. David Rusk's *Cities Without Suburbs* (1993) and Neil Peirce, Curtis Johnson, and John Stuart Hall's *Citystates* (1993) drew a good deal of public attention. Books on the subject were supported by reports from the National League of Cities (Ledebur & Barnes, 1992, 1993, 1994), big city mayors backed regional cooperation (Berkman, Brown, Goldberg, & Mijanovich, 1990), and academic researchers saw value in it (Voith, 1991, 1993; Savitch, Collins, Sanders, & Markham, 1993; Savitch, Sanders, & Collins, 1992). Much of the current rationale is rooted in propositions that (a) social and economic disparities between cities and suburbs are mounting, (b) the best way to counter local discrepancies is to join central cities and suburbs in an institutional form, and (c) economic prosperity lies in the ability of regions to act in concert, through common policies combining the strengths of cities and suburbs.

For all the common objectives, the mood and rationale of the 1990s are different from those of the preceding generation. Studies of the 1960s emphasized reducing governments for the sake of efficiency, but those of the 1990s are held together by a belief in local interdependence and economic survival. The current rationale is appealing because it furnishes a way out of central city decline through institutional change. It also holds out the hope of generating policy solutions to resolve social imbalance (desegregated housing, common investments, new infrastructure).

Realities, however, may be different. Scholars and practitioners point to the institutional and political barriers that stand in the way of metropolitan government. Some question the effectiveness of unified regional government, and a few institutions have been weakened (Barlow, 1991; Rothblatt & Sancton, 1993; Self, 1982). It is now time to look at the experience of metropolitan regions and make judgments in a systematic and comprehensive fashion. By this, we mean to examine how regionalism has evolved and the need to identify the forces that inhibit or encourage cooperation. Our concern is not so much whether we ought to have interlocal cooperation but instead to determine what the last 30 years actually have given us. What are the dividends? What are the trends?

■ Why Regions?

In many ways, metropolitan change is steered by the inconspicuous operations of regional politics. It is a type of politics rooted in political economy or, more precisely, in an interdependence between central cities and surrounding suburbs. Formally, the notion of an urban region is scarcely defined except by the Census Bureau, which classifies the nation's geography into Metropolitan Areas (MAs), Metropolitan Statistical Areas (MSAs), Primary Metropolitan Statistical Areas (PMSAs), and Consolidated Metropolitan Statistical Areas (CMSAs).[1]

Informally, regional politics consists of political networks that arise to govern clusters of localities; economic linkages that shape the growth and decline of communities; and a complex web of transportation, human habits, and social arrangements that compose America's urban sprawl. Regional politics transcends legal jurisdictions because of the need to promote economic development, protect the environment, rebuild infrastructure, deliver new services, and above all manage public policy in a competitive world. By definition, regional politics is intergovernmental, nested in economic linkages between cities and suburbs, and fueled by mobile capital, labor, and culture.

Regional politics is hardly neat, clear-cut, or explicit. The Constitution makes no mention of regions, and rarely does one see regional legislatures, chief executives, or judiciaries. With or without regional government, regions do work, though not always effectively. Decisions are made in federated patterns in which councils of government, planning bodies, and organizations of business or political elites convene. Regional economies hang together as bundles of interconnected markets. Usually, these economies are polarized toward one or more dominant centers with a high degree of vertical, horizontal, and complementary relationships (Hoover, 1975; Richardson, 1969).

Regional politics can be most acutely located at the point where business joins political power. Although one can rarely identify a formal authority at the top of this juncture, regions are held together by natural pressures for decision making and economic development. By force of these pressures, localities do interact, sometimes through stable institutions (regional councils), at other times through the offices of elected officials (governors, county executives), and often through less formal channels (development partnerships, elite networks). It is this juncture of relationships that furnishes new roles for cities and provides opportunities for collective action. Some cities have met those challenges, while others still struggle.

The much vaunted notion of cities being made obsolete by a new age of communication misses the real effects of urban change (Hawley, 1971; Sternlieb, 1971; Webber, 1963). Cities have not so much been replaced as extended. The most successful ones have melded with new forms of development. For these cities, economics, technology, and mobility have tied urban to suburban and even rural life. In many ways, cities have re-created themselves in altogether new forms, such as "edge cities" (Garreau, 1991), or produced interlocal networks through which they continue to exercise influence.

We refer to this as a post-city age and point out that regional politics has spawned a variety of arrangements designed to cope with the pressures of interdependence. We should also recognize that interdependence does not always mean cooperation, and we readily observe outcomes in which cooperation is kept within limits or even resisted. At bottom, regional politics reflects the demographic profile, the culture, and, most immediately, the political alliances and tensions of a given area. While some regions opt to promote interlocal cooperation, others favor limited engagements, and still others would prefer political divorce.

Our examination is built on the twin pillars that sustain regional politics—a region's political economy and its political institutions. By political economy, we mean the interdependence through which public and private sectors interact across local boundaries. By political institutions, we refer to the mechanisms through which regional cooperation takes place.

A region's political economy shapes its political institutions and makes certain forms of cooperation possible. Likewise, regional institutions facilitate economic vitality. This can be done by limited institutions that provide technical assistance (councils of government), it can be accomplished by functionally specific institutions in charge of mass transportation and investment (port authorities, urban development corporations), or it can be achieved by comprehensive institutions that have the power to plan, tax, and allocate (metropolitan governments). These institutions take various forms ranging from formal cooperation to loosely organized partnerships.

The next sections describe the political economy and institutional basis of 10 metropolitan regions. These are New York, Los Angeles, St. Louis, Washington, D.C., Louisville, Pittsburgh, Miami, Minneapolis-St. Paul, Jacksonville, and Portland. We begin first with political economy, comparing linkages within these regions. Following this, we turn to political institutions and develop a framework for assessing regional cooperation.

■ Interdependence Amid Disparity

The idea that cities and their surrounding areas are interdependent is not new. Jane Jacobs makes the point when she describes the etiology of economic growth. According to Jacobs (1969, 1984), city development came first, and rural development followed in its path. Without central cities, neither near suburbs nor rural hinterlands could prosper. Growth, she claims, is catalyzed when cities replace imports with internally produced goods, and, in the process, attract raw materials and labor from the countryside. The very characteristics that define a city—dense and varied population, capital, technical innovation, and markets—stimulate rural productivity and sustain a larger complex of agriculture, manufacture, trade, and transportation.

Although Jacobs may overstate the case for urban primogeniture, the synergies created by urban-suburban-rural linkages are well established. Indeed, a broad array of scholarship going back to neoclassical economic theory supports the proposition that regional economies form the nuclei of production (Heilbrun, 1987; Schumpeter, 1942; Smith, 1937; Thompson, 1972). A stream of recent studies goes further, suggesting that when central cities succeed or fail, so too do suburbs. Data on income, land values, racial segregation, and poverty show that urban and suburban fortunes are closely linked (Collins, 1994; Ledebur & Barnes, 1992, 1993; Savitch et al., 1992; Savitch et al., 1993). David Rusk (1993) claims that the keys to economic success are held by "elastic cities," capable of absorbing growing suburbs, whereas "inelastic cities," unable to annex new territory, are destined to fall into insoluble decay.

The logic is inescapable and carried by the dynamics of economic competition and the advantages of synergy. Boiled down, it can be stated as follows. Localities compete on both a national and a global scale. Industries are fed by a variety of sources, including raw materials, sophis-

TABLE 1.1   Suburban Earnings in Ten Central Counties, 1969-1989
            (thousands of 1991 dollars)

	1969	Percentage of Total	1989	Percentage of Total
New York[a]	60,658,264	68.2	72,403,884	65.5
Los Angeles	7,100,815	7.4	17,100,961	11.4
Washington, D.C.	10,605,721	54.1	16,685,902	63.1
Jacksonville	250,438	4.6	1,522,548	15.1
St. Louis	6,214,814	52.8	4,779,688	53.0
Pittsburgh	1,664,274	8.6	2,537,546	12.6
Minneapolis	1,613,737	11.5	5,271,895	22.8
Portland	2,014,461	24.8	3,554,113	33.0
Miami	1,174,794	8.5	2,577,395	10.1
Louisville	715,593	8.9	1,668,641	16.2

SOURCE: Computed from Table CA5 (Personal Income by Major Source and Earnings by Industry) by the Bureau of Economic Analysis for years shown. Values in 1991 dollars were computed by dollar amount × (1991 GDP implicit price deflator/implicit price deflator for each year). Source for GDP implicit price deflator is Survey of Current Business, September 1992, Table 3.

a. New York figures are for earnings in New York County by workers who live outside that county.

ticated transportation, a skilled labor force, research facilities, and an environment that can incubate new jobs. Standing alone, neither cities nor suburbs can provide the airports, universities, or land to harness these resources. Working together, these generative assets can be combined and coordinated to produce new products or offer something to a world that values technology, information, and managerial direction. Like it or not, therefore, localities must find ways to collaborate on policy, planning, and development.

Does the interdependence thesis hold up? Aggregate research points to a dramatic association between city and suburban well-being. Despite the findings that work is gradually decentralized, central cities still hold the highest paying jobs and help pay for the luxuries of suburban living. The higher the pay derived within a central city, the more likely its beneficiary is a suburban commuter (U.S. Department of Labor, 1976). By the same token, growth in suburbs and edge cities may stimulate downtown business districts, areas outside central cities often house back-office facilities, and these same areas furnish central cities with corporate leaders, executives, technicians, and secretaries.[2] Table 1.1 shows the amount and proportion of income earned within the central city by suburban or rural commuters and illustrates how regional economies are connected (see also the Appendix, Table A1.1).

Note that suburban incomes generated from central cities are substantial. Over half the income earned within New York, Washington, D.C., and St. Louis goes to suburbs. Minneapolis and Portland furnish upward of 20% of their payrolls to suburbanites. Even small central areas, such as Louisville and Jacksonville, contribute a meaningful portion to their regional economies. Most surprising, the percentage of income generated in the central city has gone up in 9 of the 10 cases.

The absolute amounts that flow out of central business districts into the rest of their regions is extraordinary. New York leads the pack, providing $72 billion annually to its suburban workforce. Suburbanites in Los Angeles and Washington, D.C., earn about $17 billion per year, respectively. Their counterparts elsewhere take home from $5 billion (Minneapolis-St. Paul) down to $1.6 billion (Louisville). In these times, when trillion-dollar federal deficits are taken for granted, we may be all too easily numbed by billion-dollar figures, but these are real sums, and their disappearance would have a devastating impact in city and suburb alike. Los Angeles's contribution to its region is three times more than IBM's annual earnings when that company was at its peak (Hoover, Campbell, & Spain, 1993). Minneapolis-St. Paul's contribution is equivalent to the annual amount received by America's largest recipient of foreign aid (Egypt, a nation of 55 million people). New York's contribution to its region matches the annual budget of six medium-sized states.

The image of a decaying central city surrounded by suburbs that pirate wealth from the urban core ignores the wellsprings of interdependence. The fact is cities and suburbs often grow together. Suburbs may take wealth from the urban core, but they also provide investment, profit, and a skilled labor force. Corporations know little of municipal boundaries. Instead, they straddle city and suburbs, setting up downtown headquarters while maintaining suburban back offices. Commercial decentralization, which many lament, also creates regional networks of communications, trade, and services.

These ties are evident in people's daily lives. Individuals use urban regions extensively, for work, pleasure, shopping, and social activities. Central cities are the destination of 53% of all trips taken within metropolitan areas (Pisarski, 1987). Of those who have moved to suburbs, 20% still commute to and earn their livelihoods within central cities. By now, the "reverse commute" has also become commonplace. Increasingly, people who live in central cities travel to jobs outside them—to work in suburban office towers, shopping malls, or "edge cities," or as domestic employees (see Appendix, Table A1.2).

TABLE 1.2   Commuting Patterns in Ten Metropolitan Areas, 1980-1990

	1990 Percentage of Metro Workers Commuting to Central County	Change in Percentage 1980-1990	1990 Number of Metro Workers Commuting to Central County	Percentage Change in Number 1980-1990
New York	12.87	−.24	1,406,181	2.8
Los Angeles	16.23	−.76	437,294	42.3
Washington, D.C.	23.38	−7.83	446,455	5.9
Jacksonville	41.36	4.25	45,795	104.6
St. Louis	20.66	−4.10	203,701	−2.4
Pittsburgh	24.19	2.56	87,279	11.5
Minneapolis-St. Paul	47.66	−4.97	237,509	31.1
Portland	30.48	−3.08	133,489	20.4
Miami	13.14	.07	77,285	39.3
Louisville	40.91	.00	55,445	22.2

SOURCE: Data on place of residence by place of work supplied by U.S. Bureau of the Census.

Are these patterns disappearing? Hardly. If anything, regional commutation has intensified, though the ecology of metropolitan settlement has spread. The decentralization of work has thickened an already dense network of roadways among townships, villages, "edge cities," and central cities. Commutation is not only between central cities and outlying areas but also between hamlets outside central cities. Central cities are important actors within this new ecology. Highways and trains still take a substantial portion of hinterlands into the central cities as well as out of them. Table 1.2 shows the number of workers who reside outside a central county and who work inside the central county in each of our 10 metropolitan areas (also see Appendix, Table A1.3).

Clearly, 9 out of 10 central counties gained commuters. In some cases, the gains were dramatic. Table A1.3 shows that New York County, with more than a million commuters, pumped up its inflow closer to a million and a half. Los Angeles swelled its commuter ranks to more than 400,000, while Minneapolis-St. Paul jumped to more than 200,000 commuters. Working from much smaller bases, Jacksonville doubled its commuters to 45,000, and Miami rose by nearly 40% to 77,000 daily travelers.

Skeptics might counter that cities have lost population and that increased commuter flows are composed of job holders who have since moved to the countryside. Although this is true, we should also recognize that central cities have more than held their own job bases, especially high-paying jobs. Movement out also begets movement in. Whether workers reside in the city or the country, the economic crux is that regions have become more integrated. By means of earned income, capital investment, and patterns of commutation, parts of the metropolitan area have grown together. The dissemination of jobs, common usage of airports, a shared infrastructure, and even common reference to sports teams have ramifications through a host of issues, especially those relating to politics and economics.

There is also a paradoxical side to regionalism. Regional economies may be more integrated, but their geography is more disparate. More than ever before, territories within metropolitan regions are polarized by race and class. With each passing census, the nation takes on the complexion of black or brown urban cores surrounded by white suburban peripheries. Over the last 30 years, African Americans increased their concentration in the largest 25 central cities from 20.3% to 29.8% (Collins, 1994). Racial contrasts are reinforced by economic distinctions. In 1960, every suburban dollar was matched by $1.05 in the central city. By 1990, that proportion had been reversed, plummeting to one suburban dollar for just $.84 held in the central city (Ledebur & Barnes, 1992).

Tables 1.3 and 1.4 show these disparities for 10 metropolitan areas (see also Appendix, Table A1.4). These tables include both absolute and proportional changes in regional income and racial characteristics over the last two decades. Of interest too is the ratio of city to suburban income as well as an index of residential segregation for the entire metropolitan area. This index compares the racial or ethnic mix of a given city with the mix of population in the overall metropolitan area. An index of 100 indicates that a city is totally segregated, whereas an index of 0 means that a city reflects the same proportion of minorities as the whole metropolitan area. To achieve full integration, a city with a minority index of 20 would have to move 20% of those residents into other areas.[3]

Typical of the nation at large, racial/ethnic concentrations are compounded by steep differences in city and suburban income. Of the 10 metropolitan areas, only Portland and Los Angeles enjoy an equilibrium with their suburbs, and they too show signs of slippage. Although Portland has no real minority concentration, its per capita income dropped to 92% of suburban income during the past two decades. Los Angeles still holds income parity with its suburbs, and its concentration of Hispanics is not much higher than that of its suburbs (Appendix, Table A1.4). St. Louis appears to be the most disparate, showing a high proportion (47%) of

TABLE 1.3 Racial Concentration in Ten Metropolitan Areas, 1970 and 1990

Metropolitan Area	Percentage Black 1970		Percentage Black 1990		Residential Segregation (Black/White) 1990
	City	Suburbs	City	Suburbs	
New York	21.1	6.0	28.7	12.1	83
Los Angeles	17.9	6.1	14.0	9.4	74
Washington, D.C.	71.1	8.3	65.8	19.4	67
Jacksonville	22.3	17.5	25.2	7.7	64
St. Louis	40.9	7.0	47.5	11.5	80
Pittsburgh	20.2	3.5	25.8	4.3	n.a.
Minneapolis-St. Paul	4.0	.2	10.7	1.2	n.a.
Portland	5.6	.3	7.7	.6	68
Miami	22.7	12.2	27.4	19.0	72
Louisville	12.8	3.3	29.7	6.6	73

SOURCE: *County and City Data Book 1977,* Table 4; *State and Metropolitan Area Data Book, 1979,* Table B; Slater, C., and Hall, G. (Eds.). (1993). Lanham, MD: Burnan Press. *1996 County and City Extra;* 1991 *USA Today* (November 11, 1991, p. 4) for residential segregation indices.

TABLE 1.4 Central City and Suburban[a] Per Capita Income in Ten Metropolitan Areas, 1969 and 1989

	1969[b]			1989		
	Central City	Suburban	Ratio	Central City	Suburban	Ratio
New York	3,698	4,293	.86	16,281	24,072	.68
Los Angeles	3,951	3,806	1.04	16,188	16,124	1.00
Washington, D.C.	3,842	4,406	.87	18,881	21,880	.86
Jacksonville	2,853	2,853	—	13,661	15,264	.89
St. Louis	2,726	3,498	.78	10,798	15,715	.69
Pittsburgh	3,071	3,171	.97	12,580	14,375	.88
Minneapolis	3,483	3,653	.95	14,830	17,196	.86
Portland	3,533	3,456	1.02	14,478	15,726	.92
Miami	2,821	3,647	.77	9,799	14,569	.67
Louisville	2,958	3,269	.90	11,527	14,416	.80

SOURCE: For 1969, *County and City Data Book 1972;* for 1979, *1986 State and Metropolitan Area Data Book;* for 1989, Census Summary Tape File 3C.
NOTE: Jacksonville was a one-county SMSA in 1969, so city and SMSA boundaries were coterminous.
a. Suburban figures are for remainders of PMSAs in 1989 and remainders of SMSAs in 1969.
b. 1969 figures are derived from per capita money income comparisons.

African Americans while its suburbs held just 11.5%. Most alarming, St. Louis holds just 69% of suburban income, down from an already low proportion two decades ago. With 80% of its black residents living in concentrated areas, St. Louis's segregation is among the highest in the nation. Miami and New York also have a pattern of racial imbalance together with lower income ratios for their suburbs. By 1989, Miami and New York income fell below .70 to the suburban dollar. Although Washington, D.C., is one of the most racially concentrated and segregated cities in America, its income ratio is not bad and has remained steady over the last 20 years. This is undoubtedly a result of the city's position as the nation's capital and its abundance of public sector jobs.

What can we make of these mirror images of interdependence and disparity? For one, they reflect tensions that are endemic to metropolitan areas. Cities and suburbs need each other but are also uneasy about each other. The urban attractions of jobs and pleasure are offset by a suburban fear of crime, crowding, and incipient decay. The suburban attractions of green space and comfortable living are countered by a city sense that the hinterlands are racist, exploitive, and dull. Second, and in response to these perceptions, cities and suburbs adopt self-enhancing measures. Given the chance, cities tax suburban commuters with levies on payrolls, hotel occupancy, office rentals, and sales. Suburbs limit their land to the most affluent urbanites through restrictive zoning and high prices. Last, these forces of attraction (interdependence) and repulsion (disparity) are refracted in regional politics. They can be spotted most vividly in how a metropolis organizes its institutions and conducts its business. Those regions beset by severe disparities are apt to treat their localities warily and feel their way through incremental or limited cooperation. Those regions less stricken by inequalities are more inclined toward more comprehensive, formalized relationships.

■ Patterns of Institutional Refraction

The 1960s critique of local government still holds. Throngs of local jurisdictions blanket the country, including 282 Metropolitan Statistical Areas and more than 33,000 local authorities (U.S. Department of Commerce, Bureau of the Census, 1992). This involves counties, municipalities, townships, and various kinds of special districts. The average metropolitan area contains roughly 117 governing units, making America one of the most fragmented nations on Earth. Municipalities, which are the most politically potent units, have proliferated during the last few decades (up 20%) and continue to multiply.

Although some scholars lament this trend and argue for consolidation, public choice theorists see virtue in fragmentation. The debate between consolidators and public choice theorists concerns questions of efficiency, cost, service delivery, economic growth, and social reform. Consolidators come from a long reformist tradition and seek to rationalize government, make it understandable to citizens, and redistribute public resources (Committee for Economic Development, 1996; Frisken, 1991; Lyons & Lowery, 1989; Newton, 1975; Phares, 1989; Rusk, 1993; Wood, 1961). They claim that consolidated jurisdictions provide economies of scale, enable localities to pool resources for development, and allow citizens to participate more fully in government. Rusk (1993) makes his case for consolidated government on the grounds that it furthers racial integration.

On the other side, public choice theorists see metropolitan regions as a vast public market in which citizens choose between contending public providers (Ostrom, Tiebout, & Warren, 1961; Parks & Oakerson, 1989; Teaford, 1979). They also maintain that interlocal rivalry reduces costs and makes governments more efficient (Parks & Oakerson, 1989; Schneider, 1986). Some public choice theorists reject the notion that metropolitan regions are fragmented, preferring instead to see a "complex local political economy" offering a diversity of services (Parks & Oakerson, 1989).

The debate has consumed mounds of paper, with each side offering its own solutions and rationale. There may be another way to view the organization of metropolitan regions, not as occurring through rational deliberations about the "best form of government" nor through the invisible hand of an efficient market place. Rather, the propulsion toward metropolitan organization is influenced by interdependence, which presents economic opportunities for business as well as political possibilities for politicians and bureaucrats. History suggests that cities annex because it augments the tax base and boosts local prestige (Fleischmann, 1986; Kotler, 1969; Marando, 1979). Politicians also see a chance to enlarge their constituencies (more people mean more prestige, larger campaign donations, visibility, and higher office), and bureaucrats see the prospects of greater control (more resources, greater territorial scope). What stops or modifies the process is resistance, stemming from disparity—either by affluent suburbs that defensively incorporate or by the sheer demographic and racial disharmony of a region.

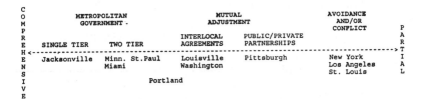

Figure 1.1. A Continuum of Regional Institutions

To help in analyzing the outcome of these dynamics, we identify some responses to regional pressures. First is formal metropolitan government within a region (Barlow, 1991), which can take the form of a single-tier unification or two-tier metropolitan or federated government. Second is a process of mutual adjustment, which can take the form of interlocal agreements among municipalities, counties, special districts, and regional authorities within a metropolitan area or region (Rothblatt & Sancton, 1993) or, alternatively, public-private partnerships among business, government(s), and citizens in a metropolitan area or region (Peirce et al., 1993). In the case of interlocal agreements and public-private partnerships, coordination comes about because of informal and formal accords among local governments and private actors. In most cases, these arrangements do not yield formal institutions of metropolitan government but are carried out by existing agencies or networks of actors.

Although the emphasis here is on how institutions promote regional cooperation, we also should recognize that regional cooperation may often be the exception rather than the rule. The response to pressures for regional governance may also include avoidance and conflict. These may occur sporadically or may be predominant responses. Avoidance and conflict, then, characterize a third basic type of response to regional pressures. Figure 1.1 places each of these types on a continuum, ranging from the most comprehensive form of regional cooperation to those that are partial and finally to noncooperation, labeled avoidance and conflict.

Metropolitan government most closely approximates the ideal of consolidators, who see comprehensive government as a solution to metropolitan problems and the best way to achieve efficiency. Consolidators also believe that metropolitan government can more effectively promote economic development, reduce fiscal inequality, and deliver services across jurisdictional lines. These advocates point with pride to Miami-Dade, Minneapolis-St. Paul, and consolidated city-counties such as Jacksonville.[4]

On the other side of the question, public choice scholars point to St. Louis and Pittsburgh as examples of "complex local organization" that affords citizen-consumers greater choice. These choices are embodied in interlocal agreements and public-private partnerships. All of this is made possible through a complex process of mutual adjustment among localities as well as between public and private sectors within regions (ACIR, 1993).

Avoidance and conflict are inadvertent or spontaneous responses to regional pressures. Although few politicians may be willing to admit to such a response, they are not uncommon and account for three of our cases. More often than not, these responses are couched in racial, social, or class differences, reinforced by discrimination and patterns of settlement. Because of this, avoidance and conflict is more common in older industrial areas (St. Louis) rather than newer, less densely packed regions (Portland).

The cases in this volume are chosen for their distinct approaches to regional pressures. We include four examples of metropolitan government; a one-tier case (Jacksonville); and three two-tier cases (Minneapolis-St. Paul, Portland, and Miami). We also examine mutual adjustment cases involving interlocal cooperation (Louisville and Washington, D.C.) and public-private partnerships (Pittsburgh). Last, we point up the predominance of avoidance and conflict (Los Angeles, New York, and St. Louis). Table 1.5 provides a composite illustration of the regions, their approach to governance, their definition, and their size.

These are ideal types, categorizations that enable us to identify essential features. The realities of regional institutions are sometimes hybrids and can even incorporate uncategorized features. At still other times, one type may begin to blend with another. Matching the concepts against actual arrangements poses some vexing questions. When, for example, do extensive mutual adjustments begin to resemble two-tier governance? How do we classify a once vibrant two-tier government, now undercut by politicians and forced to emphasize its role as a regional advisory council? Most important, how do we balance a limited number of mutual adjustments against a backdrop of conflict and avoidance? Even the most conflict-ridden regions possess some form of regional cooperation (usually limited to functional agencies such as port authorities and mass transit operations). For us, the determination rests on which pattern is most

TABLE 1.5 Approaches to Regional Governance

Case	Approach to Regional Governance	Functional City	Population Size, 1990
New York City Bruce Berg and Paul Kantor	Avoidance and Conflict	City of New York City and Counties of New York, Kings, Richmond, Queens, Bronx, Westchester, Nassau, Rockland, Orange, Putnam, and Suffolk, NY; Hudson, Bergen, Essex, Passaic, Union, Hunterdon, Middlesex, Monmouth, Morris, Ocean, Somerset, and Sussex, NJ; and Fairfield, CN	18,093,000
Los Angeles Alan L. Saltzstein	Avoidance and Conflict	City of Los Angeles and Counties of Los Angeles, Orange, Riverside, San Bernardino, and Ventura, CA	14,530,000
St. Louis Donald Phares and Claude Louishomme	Avoidance and Conflict	City of St. Louis and County of St. Louis, MO	1,390,214
Washington, D.C. Jeffrey Henig, David Brunori, and Mark Ebert	Mutual Adjustment: Intergovernmental Cooperation	Cities of Washington, D.C., and Alexandria, Fairfax, Falls Church, Manassas, and Manassas Park, VA, and the Counties of Calvert, Charles, Frederick, Montgomery, and Prince George, MD; and Arlington, Loudoun, Fairfax, and Prince William, VA	3,900,000
Louisville H. V. Savitch and Ronald K. Vogel	Mutual Adjustment: Intergovernmental Cooperation and Public-Private Partnership	City of Louisville and Counties of Jefferson, Oldham, Shelby, and Bullitt, KY; and Clark, Floyd, Harrison, and Scott, IN	973,653
Pittsburgh Louise Jezierski	Mutual Adjustment: Public-Private Partnership	City of Pittsburgh and Counties of Allegheny, Beaver, Fayette, Washington, and Westmoreland, PA	2,056,705
Miami Genie Stowers	Metropolitan Government (two-tier)	City of Miami and County of Dade, FL	1,982,900
Minneapolis-St. Paul John Harrigan	Metropolitan Government (two-tier)	Cities of Minneapolis-St. Paul and Counties of Hennepin, Ramsey, Washington, Dakota, Anoka, Scott, and Carver, MN	2,464,000
Jacksonville Bert Swanson	Metropolitan Government (one-tier)	City of Jacksonville and Counties of Duval, Clay, Nassau, and St. Johns, FL	906,727
Portland Arthur Nelson	Metropolitan Government (two-tier)	City of Portland and Counties of Clackamas, Multnomah, and Washington, OR	1,174,291

15

predominant, and we categorize regions according to their principal characteristics.

■ The Functional City and Regional Politics

Although traditional American cities are in trouble, they are not obsolete. In many ways, they are vibrant: Cities have changed by extending themselves and by interacting with areas around the urban core. We call this region the functional city and in the following chapters illustrate its patterns of development, its political institutions, and the ebb and flow of regional cooperation and conflict. We also show how economic and social pressures combine to create regional politics. Part of this analysis reminds us that political institutions reflect existing tensions. The struggle to build institutions and formulate solutions in the light of those tensions explains why regions behave differently.

Finally, we note how definitions of a metropolitan region or "functional city" may be variable. The degree to which metropolitan or regional governance exists in the United States is an artifact of the operational definition of a city-region. If we define city-regions as corresponding with counties, then metropolitan governance exists virtually everywhere. A more accurate definition of a city-region might rely on the MSA as a regional entity, but this poses other problems. Some MSAs run across state lines, MSA boundaries are not often coterminous with political institutions, and MSAs do not reflect regional community identities. To be sure, any unit of analysis we select will have advantages and disadvantages. In weighing these factors, we elected to take a more flexible and adaptive approach to this issue. Our decision was to tailor regional definitions to locally relevant conditions and criteria. Each author's expertise was brought to bear in establishing a regional definition, and the rationale is explained within each chapter. Although not offering fixed and neatly drawn definitions, this approach ensures that regions are a product of local meaning, connected to the realities of economic intercourse, social identities, and political institutions.

APPENDIX

The following shows each city's population as a percentage of its central county population and land area. Note that New York comprises five counties, that Washington, D.C., comprises the District of Columbia, Jacksonville is coterminous with Duval County, and St. Louis is a separate entity from St. Louis County. The figure for Minneapolis-St. Paul reflects the combined populations (and areas) of the two component cities as a percentage of the combined populations (and areas) of each of their central counties.

TABLE A1.1 Percentage of Central County

City	Population	Land Area	City Area (km^2)	County Area (km^2)
New York	Coterminous with 5 counties	Coterminous with 5 counties	800	800
Los Angeles	39	12	1,216	10,515
Washington, D.C.	Coterminous	Coterminous	159	159
Jacksonville	Coterminous	Coterminous	2,004	2,004
St. Louis	Coterminous	Coterminous	160	n.a.
Pittsburgh	28	8	144	1,891
Minneapolis-St. Paul	42	15	279	1,846
Portland	75	29	323	1,127
Miami	19	2	92	5,036
Louisville	40	16	161	997

SOURCE: Slater, C., and Hall, G. (Eds.). (1993). *1993 County and City Extra*. Lanham, MD: Burnan Press.

The reverse commutation figures for our 10 regions are worth noting and show a significant increase over the last decade.

TABLE A1.2   Reverse Commutation, 1980-1990 (number of workers living in central county working outside central county)

	1980[a]	1990	Percentage Change
New York County	103,503	118,397	14.39
Los Angeles	138,245	242,938	75.73
Washington, D.C.	56,510	67,694	19.79
Jacksonville	14,306	18,284	27.81
St. Louis	41,386	54,318	31.25
Pittsburgh	33,378	39,639	18.76
Minneapolis-St. Paul	43,679	75,026	71.77
Portland	32,358	54,834	69.46
Miami	27,243	43,274	58.84
Louisville	14,327	20,772	44.98

SOURCE: Census Bureau tables showing commuter flows.

a. The 1980 figures are estimated; for example, for New York County, 91,470 people actually reported that they worked outside of the county. This was 15.23% of the workers in that county. Not all workers in 1980 reported their place of work, however, including 79,224 of the 679,599 workers living in New York County. We therefore added 15.23% of those who did not report their place of work to the totals for 1980.

Following are the commutation figures in tabular form, showing the number of people who work inside the central county and live in a given metropolitan area but outside the central county.

TABLE A1.3 Commuting Patterns 1980-1990

	1980[a]	1990	*Percentage Change*
New York County	1,367,325	1,406,181	2.8
Los Angeles	307,324	437,294	42.3
Washington, D.C.	421,762	446,455	5.9
Jacksonville	22,379	45,795	104.6
St. Louis	208,745	203,701	−2.4
Pittsburgh	78,300	87,279	11.5
Minneapolis-St. Paul	181,234	237,509	31.1
Portland	110,872	133,489	20.4
Miami	55,486	77,285	39.3
Louisville	45,378	55,445	22.2

SOURCE: Census Bureau tables showing commuter flows.

a. The 1980 figures are estimated; for example, in 1980, 1,236,973 people residing outside New York County but within the CMSA reported working in New York County (this represented 19.1% of the people in that area). However, 683,904 people in that area did not report their place of work, so we added the same percentage (19.1%) of those who did not report their place of work to the total for New York County (assuming that the same percentage of those persons would have worked in New York County).

TABLE A1.4 Hispanics

Metropolitan Area	*Percentage Hispanic 1970*		*Percentage Hispanic 1990*		*Residential Segregation (Hispanic/White) 1990*
	City	*Suburbs*	*City*	*Suburbs*	
New York	10.3	21.0	24.4	8.7	68
Los Angeles	18.4	12.6	40.0	36.4	63
Washington, D.C.	2.1	1.8	5.4	5.8	43
Jacksonville	1.3	3.1	2.6	2.2	25
St. Louis	1.0	1.6	1.3	1.0	29
Pittsburgh	*	*	.9	.5	n.a.
Minneapolis-St. Paul	1.4	1.4	2.8	.9	n.a.
Portland	1.7	.4	3.2	3.8	31
Miami	45.3	13.8	62.5	46.2	52
Louisville	.4	3.0	.7	.6	34

SOURCES: *County and City Data Book 1977,* Table 4; *State and Metropolitan Area Data Book, 1979,* Table B; Slater, C., and Hall, G. (Eds.). (1993). *1993 County and City Extra.* Lanham, MD: Burnan Press. 1991 USA Today for residential segregation indices.

*Exact numbers of Hispanics in the city of Pittsburgh not given for 1970 (less than 400).

NOTES

1. MAs (metropolitan areas) as defined by the Census Bureau generally contain at least one central city (minimum population of 50,000) and comprise one or more central counties. An MA may also include outlying counties. Within this definition, MSAs are stand-alone MAs that are surrounded by nonmetropolitan counties. If an MA has more than one million persons, however, primary metropolitan statistical areas (PMSAs) may be defined within it. A PMSA includes a large urbanized central county or counties with strong internal (economic and social) links as well as close ties to other portions of the larger area. When a PMSA is defined, the larger area of which it is a component is designated as a consolidated metropolitan statistical area (CMSA).

2. The point here is not to assert suburban dependence on central cities but interdependence among jurisdictions. In some cases (Los Angeles), early growth in suburban areas outpaced that in the central city and later contributed to the rise of downtown business districts. Thus, growth and vitality can come from more than one direction (central city to suburb as well as suburb to central city). The relationship between cities and their outlying areas is often reciprocal. For details on this question, consult the *Urban Affairs Review,* November 1995.

3. Residential segregation is usually measured by **D** (the index of dissimilarity), which is computed as one-half the sum of the absolute differences (positive and negative) between the percentage distributions of the black or Hispanic population and the white population in a metropolitan area (census tracts compared to entire area). Whatever the value, it shows the percentage of either group that would have to move from one tract to another to achieve an even spatial distribution throughout the metropolitan area. For more information, see Darden (1990).

4. Other consolidations that are not the subject of our case studies include Lexington, Kentucky; Indianapolis, Indiana; and Nashville, Tennessee.

REFERENCES

Advisory Commission on Intergovernmental Relations. (1967). *Fiscal balance in the American federal system* (Vol. 2). Washington, DC: Author.

Advisory Commission on Intergovernmental Relations. (1969). *Urban America and the federal system.* Washington, DC: Author.

Advisory Commission on Intergovernmental Relations. (1993). *Metropolitan organizations: Comparison of the Allegheny and St. Louis case studies.* Washington, DC: Author.

Aron, J. B. (1969). *The quest for regional cooperation: A study of the New York Metropolitan Regional Council.* Berkeley: University of California Press.

Barlow, I. M. (1991). *Metropolitan government.* New York: Routledge.

Berkman, R., Brown, J. F., Goldberg, B., & Mijanovich, T. (Eds.). (1992). *In the national interest: The 1990 urban summit: With related analysis transcript, and papers.* New York: Twentieth Century Fund Press.

Collins, D. A. (1994). *Central city governmental inclusion and city-suburban disparities.* Unpublished doctoral dissertation, Urban and Public Affairs, University of Louisville.

Committee for Economic Development. (1967). *Modernizing state government.* New York: Author.

Darden, J. T. (1990, July). *Residential segregation of blacks and Hispanics: The significance of race, ethnicity, and class.* Paper presented at the International Housing Research Conference, Paris.

Fleischmann, A. (1986). The goals and strategies of local boundary changes: Governmental organization or private gain? *Journal of Urban Affairs, 8*(4), 63-76.

Frisken, F. (1991). The contributions of metropolitan government to the success of Toronto's public transit system: An empirical dissent from the public-choice paradigm. *Urban Affairs Quarterly, 27,* 268-292.

Garreau, J. (1991). *Edge city: Life on the new frontier.* New York: Anchor.

Hawley, A. H. (1971). *Urban society: An ecological approach.* New York: Ronald.

Heilbrun, J. (1987). *Urban economics and public policy* (3rd ed.). New York: St. Martin's.

Hoover, E. M. (1975). *An introduction to regional economics.* New York: Knopf.

Hoover, G., Campbell, A., & Spain, P. J. (1993). *Hoover's handbook of American business.* Austin, TX: Reference Press.

Jacobs, J. (1969). *The economy of cities.* New York: Vintage.

Jacobs, J. (1984). *Cities and the wealth of nations.* New York: Vintage.

Kotler, M. (1969). *Neighborhood government: The local foundations of political life.* Indianapolis: Bobbs-Merrill.

League of Women Voters Education Fund (1974). *Supercity/hometown, U.S.A.: Prospects for two-tier government.* New York: Praeger.

Ledebur, L. C., & Barnes, W. R. (1992). *Metropolitan disparities and economic growth.* Washington, DC: National League of Cities.

Ledebur, L. C., & Barnes, W. R. (1993). *"All in it together": Cities, suburbs and local economic regions.* Washington, DC: National League of Cities.

Ledebur, L. C., & Barnes, W. R. (1994). *Local economies: The U.S. common market of local regions.* Washington, DC: National League of Cities.

Lyons, W. E., & Lowery, D. (1989). Governmental fragmentation versus consolidation: Five public-choice myths about how to create informed, involved, and happy citizens. *Public Administration Review, 49,* 533-543.

Marando, V. L. (1979). City-county consolidation: Reform, regionalism, referenda, and requiem. *Western Political Quarterly, 32,* 409-421.

Newton, K. (1975). American urban politics: Social class, political structure and public goods. *Urban Affairs Quarterly, 11,* 241-264.

Ostrom, V., Tiebout, C., & Warren, R. (1961). The organization of government in metropolitan areas: A theoretical inquiry. *American Political Science Review, 55,* 831-842.

Parks, R. B., & Oakerson, R. J. (1989). Metropolitan organization and governance—A local public economy approach. *Urban Affairs Quarterly, 25*(1), 18-29.

Peirce, N. R., Johnson, C. W., & Hall, J. S. (1993). *Citystates: How urban America can prosper in a competitive world.* Washington, DC: Seven Locks.

Phares, D. (1989). Bigger is better, or is it smaller?: Restructuring local government in the St. Louis area. *Urban Affairs Quarterly, 25,* 5-17.

Pisarski, A. E. (1987). *Commuting in America.* Westport, CT: Eno Foundation for Transportation.

Richardson, H. W. (1969). *Regional economics.* New York: Praeger.

Rothblatt, D. N., & Sancton, A. (Eds.). (1993). *Metropolitan governance: American/Canadian intergovernmental perspectives.* Berkeley: Institute of Governmental Studies Press.

Rusk, D. (1993). *Cities without suburbs*. Washington, DC: Woodrow Wilson Center Press.

Savitch, H. V., Collins, D., Sanders, D., & Markham, J. P. (1993). Ties that bind: Central cities, suburbs, and the new metropolitan region. *Economic Development Quarterly, 7*, 341-357.

Savitch, H. V., Sanders, D., & Collins, D. (1992). The regional city and public partnerships. In R. Berkman, J. F. Brown, B. Goldberg, & T. Mijanovich (Eds.), *In the national interest: The 1990 urban summit: With related analysis transcript, and papers.* New York: Twentieth Century Fund Press.

Schneider, M. (1986). Fragmentation and the growth of local government. *Public Choice, 48*, 255-263.

Schumpeter, J. A. (1942). *Capitalism, socialism and democracy*. New York: Harper & Row.

Self, P. (1982). *Planning the urban region: A comparative study of policies and organizations.* University: University of Alabama Press.

Slater, C., & Hall, G. (Eds.). (1993). *1993 county and city extra*. Lanham, MD: Barnan.

Smith, A. (1937). *An inquiry into the nature and causes of the wealth of nations.* New York: Modern Library.

Sternlieb, G. (1971). The city as sandbox. *The Public Interest, 25*(Fall), 14-21.

Teaford, J. C. (1979). *City and suburb: The political fragmentation of metropolitan America, 1850-1970.* Baltimore: Johns Hopkins University Press.

Thompson, W. R. (1972). *A preface to urban economics*. Baltimore: Johns Hopkins University Press.

U.S. Department of Commerce, Bureau of the Census (1992). *Government organization, census of governments, volume 1, number 1.* Washington, DC: Author.

U.S. Department of Labor, Bureau of Labor Statistics, New York, May 20, 1976.

Voith, R. P. (1991). Capitalization of local and regional attributes into wages and rents: Differences across residential, commercial and mixed use communities. *Journal of Regional Science, 31*, 127-145.

Voith, R. P. (1992). *Do declining cities hurt their suburbs?* Unpublished manuscript.

Voith, R. P. (1993). Changing capitalization of CBD-oriented transportation systems: Evidence from Philadelphia, 1970-1988. *Journal of Urban Economics, 33*(3), 361-376.

Webber, M. (1963). Order in diversity: Community without propinquity. In L. Wingo (Ed.), *Cities and space: The future use of land.* Baltimore: Johns Hopkins University Press.

Wood, R. (1961). *1400 governments*. Cambridge, MA: Harvard University Press.

Part I

Avoidance and Conflict

2

New York:
The Politics of
Conflict and Avoidance

BRUCE BERG
PAUL KANTOR

■ **Introduction**

This study reexamines the nation's largest metropolitan area, the New York region, and finds the last 20 years dominated by a politics of conflict and competition. In a political landscape of fragmentation and rivalry among governments, the region's political entrepreneurs have found few opportunities for forging regional solutions to public problems. Further, postindustrial restructuring of the region's economy is unleashing powerful new forms of intergovernmental economic competition. This new reality seems likely to further diminish chances for greater regional cooperation in the future.

For our purposes, the New York region comprises the 24 contiguous counties in three states that together form the New York–Northern New Jersey–Long Island Consolidated Metropolitan Statistical Area (CMSA). Eleven of these counties are in New York (the five counties of New York City and six suburban counties), 12 are in Northern New Jersey, and 1 is in Connecticut. This is one of the largest urban concentrations in the world and comprises a population of 18,093,000.[1]

Robert Wood called the tristate region centering on New York City "one of the great unnatural wonders of the world" (Wood, 1961, p. 1). This giant

AUTHORS' NOTE: We wish to thank Joseph Trapanese and Heather Wright Feinsilver, doctoral students in the Political Science Department at Fordham University, for assisting in data collection in preparation of this work. The Fordham University Research Council provided financial assistance.

displays considerable economic interdependence (Hoover & Vernon, 1959). With its soaring skyline, Manhattan visibly plays a powerful centripetal role as a central business district (CBD). Although only about 8% of the region's population lives in Manhattan, the Manhattan CBD draws more than 41% of commuters on a daily basis (Port Authority of New York and New Jersey, 1994b, p. 30). A center of national and international commerce and industry, New York City below 60th Street constitutes a powerful economic engine, the performance of which helps drive the economy of the region as a whole.

Transportation interdependence also gives the region a huge capacity for moving goods, people, and information. The New York region is held together by a complex of five airports, bistate seaports, the nation's largest mass transit system, a busy commuter rail, and thousands of miles of highway that are scattered all over the 24-county area. The telecommunications system is the standard for the world. There are more computers in Manhattan than in all of Europe and more "smart" office buildings (equipped with optical fiber technology and advanced computer and telecommunications systems) than in any other city in the world (Lloyd & Kahn, 1991, p. 37).

Development as an international center in finance, business services, media, and culture also ties city and suburb to essentially similar investment flows in the global marketplace. For example, the region has 450 offices of foreign banks, which hold more than $603 billion in assets, or more than two-thirds of all foreign banking assets in the United States. The region also has 35% of the nation's security jobs and more than 20% of America's advertising jobs (Port Authority of New York and New Jersey, 1994a, pp. 4-5).

Nevertheless, this remarkable region has flourished without great governmental planning (Danielson & Doig, 1982, pp. 30-31). It is among urban America's most fragmented political areas. The region constitutes the intersection of three states having essentially different political traditions and political characteristics. New Jersey lacks many large cities and is dominated by suburban political views. Connecticut has New England affinities, and only a small part of the state is within the tristate region. New York City's political importance looms exceptionally large in Albany, but not in the other two state capitals.

Although New York City, with its more than 7 million citizens and its network of five counties, dominates the governmental scene in respect to size, the 1992 census of governments places it among a total of 1,787 county, municipal, town, school district, and special district governments. This number has grown since 1977, when 1,776 jurisdictions were counted (Bureau of the Census, 1992). This accounting actually under-states the degree of governmental fragmentation. It does not include all public authorities (called public benefit corporations in New York State), and such entities have been created in increasing numbers during recent decades (Kantor, 1995; Walsh, 1991), nor does this include the state and federal housing, transportation, and other departments that have program responsibilities in the region.

These numerous local governmental institutions in the region all share a strong tradition of home rule that is respected in their state capitals. Consequently, these governments dominate land use, education, planning, housing, and economic development policies, and they deliver the impor-tant services besides transportation. Unlike some other metropolitan ar-eas, New York lacks a single general-purpose governmental institution with a responsibility for the entire region. The agency with the widest scope is the Port Authority of New York and New Jersey (PA), which was created by compact between the two states in 1921. Today, it operates much of the region's transportation infrastructure, including airports, bridges, tunnels, commuter railroads, and seaport facilities.

The Metropolitan Transit Authority (MTA) is the largest mass trans-portation conglomerate in the United States. Its operating agencies run the New York City bus and subway systems, commuter railroads in the northern and eastern counties, a metropolitan bus system, and the revenue-generating Tri-borough Bridge and Tunnel Authority that once was Robert Moses' power base. The MTA's services are limited to the New York side of the region and part of Connecticut. All other agencies have more limited scope and purpose. As explained later, private sector regional planning agencies have not been able to step into the breach left by the public planners.

■ The Political Economy of Regional Change

Regional political relationships may change, however. Political co-operation may grow as a result of new common interests that give everyone important reasons to pull together or face punishing losses in income and power. Support for cooperative solutions may arise as a result of the decline of old economic and social divisions that are a source of political conflicts. It is equally possible that political cooperation may diminish if interest-formation processes work in the opposite directions, giving local officials new reasons for intergovernmental rivalry.

In recent decades, the region's political tradition of conflict and avoidance has been put to the test by postindustrial economic change. Technological breakthroughs in communication, production, and transportation have permitted greater dispersal of business activities. Postindustrial development also has altered the distribution of job activities (Beauregard, 1991; Kantor, 1995). Manufacturing activities have shrunk as these industries have become less competitive in world markets, while service employment has grown as a larger component of urban economies almost everywhere.

The political consequences of postindustrial restructuring on regional governmental relations cut two ways. On one hand, as business units become more mobile, governments in the same metropolitan area are constrained to compete more as a region. This enhances shared interests in intraregional political cooperation, at least on economic matters, because business location decisions are apt to be made on the basis of the business services, labor supply, housing, transportation, and communications facilities available in the region as a whole, not only the immediate political jurisdiction where business is located. On the other hand, economic changes also can diminish interest in regional political cooperation. If restructuring favors some parts of the region at the expense of other parts, greater economic competition to capture essentially similar economic roles weakens shared intergovernmental economic interests. Thus, it is possible for postindustrialism simultaneously to have both effects on the region's political jurisdictions. Nevertheless, the stimulation of new political interests that foster cooperation seems greater if economic restructuring avoids forms of uneven regional development that stimulate economic rivalry. The more that some communities capture the jobs and functions at the expense of other communities within the same region, conflict, rather than cooperation, is likely to grow.

The sharing of common economic interests is undoubtedly a necessary but not a sufficient condition for the growth of regional political cooperation. Ultimately, regional political cooperation requires skillful political entrepreneurship. Economic and social changes influence the character of political interests, increasing or diminishing opportunities for intergovernmental cooperation—but political cooperation does not simply happen (Sanders & Stone, 1987; Stone, 1989). In a system of dispersed power, political cooperation must be created by skillful political power brokers.

These leaders must set new agendas and build political coalitions that can support regional solutions to specific problems.

Thus, change in the political economy of regional cooperation can be conceived as a dual process. Postindustrial changes create new common intergovernmental interests in regional affairs. These changing interests are, in turn, acted on by political entrepreneurs who seek to build regional political coalitions that overcome the fragmented political character of the system. The following pages examine both of these processes in the New York region during the last several decades. We first analyze the character of postindustrial economic change in the region, highlighting interest-formation impacts. The next section examines the political responses to such change by major political entrepreneurs and institutions.

■ The Postindustrial Economy and Regional Interests

For analytical purposes, the New York CMSA is best looked at by comparing county political jurisdictions that share essentially similar histories in respect to age of development, physical infrastructure, and population density or that are bound together as a result of physical proximity to the area's historic urban core, Manhattan. Accordingly, we treat the New York CMSA counties as three successive rings that surround the Manhattan CBD (Figure 2.1).

The "core counties" of Ring 1 comprise the four outer borough counties of New York City (Brooklyn [Kings], Queens, the Bronx, and Richmond), as well as nearby Hudson County in New Jersey.[2] This groups together the counties adjacent to the CBD that were developed mainly during the 19th century and the 1920s, along with the only part of New York City (Staten Island) that was developed more recently. The "inner suburbs" of Ring 2 were settled mostly during the interwar and immediate postwar years. The "outer suburbs" of Ring 3 are those jurisdictions that developed largely since World War II and have experienced significant population growth during the last several decades.[3]

Two major forms of postindustrial economic change have impacted the region since the 1970s. One is the restructuring of the economy as old industries fade and new ones, especially service-related industries, have grown. The other is a trend toward steady, if not rapid, economic growth in the suburbs and declining or stagnant growth in the CBD and its inner ring.[4] As a result, the CBD's share of the metropolitan economy is

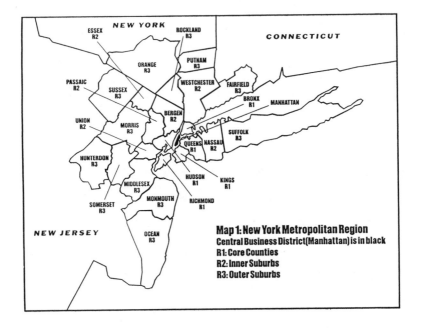

Figure 2.1.   Map 1: New York Metropolitan Region

declining while the suburban share grows. Both of these trends are diminishing the traditional economic importance of the CBD while encouraging the rise of new suburban agglomeration economies that replicate and compete with the CBD. "New suburbs" have evolved from bedroom suburbs, dependent upon the CBD for the bulk of their economic activity, into "outer cities" that have achieved a degree of independence from the CBD (Stanback, 1991, p. 60). While regional economic interdependence continues, the uneven and competitive character of this type of regional development does not augur well for political cooperation. On balance, it represents the formation of new rival economic interests between city and suburb as well as among suburbs.

TABLE 2.1 Employment in the New York Metropolitan Region, 1970-1990

	Percentage Change in Employment		Percentage of Region Total		
	1970-1980	*1980-1990*	*1970*	*1980*	*1990*
Total region	3.3	15.3			
CBD	−11.1	7.1	32.5	27.9	25.9
Core counties	10.6	10.5	21.1	18.2	17.5
Inner suburbs	8.4	11.6	27.5	28.8	27.9
Outer suburbs	36.8	32.5	18.7	24.8	28.5

SOURCE: Data taken from Regional Economic Information System, Bureau of Economic Analysis, U.S. Department of Commerce.

A look at overall employment changes conveys an image of an urban core whose growth is slow, surrounded by more dynamic and prosperous suburban counties. Employment growth in the region strongly favored the suburban rings, as suggested in Table 2.1. As a percentage of the entire region, total employment in the CBD and the core counties declined between 1970 and 1990 from 53.6% to 43.4%. This decline in the share of the regional employment base occurred even though both the CBD and the core counties experienced positive job growth during the 1980s. Total employment in the inner suburbs as a percentage of the entire region was stable over the two decades. The outer suburb's share of regional total employment base increased significantly, going from 18.7% to 28.5%. By 1990, the inner and outer suburbs contained 56.4% of the region's employment, an increase of more than 10% since 1970.

Changes in earnings by place of work convey a similar picture of suburban economies that are differentiating and maturing. Table 2.2 displays income earned by place of work for the subdivisions of the region. The table includes the rate of growth of income earned by place of work for each of the region's subdivisions as well as each subdivision's percentage of total regional income earned by place of work for 1970, 1980, and 1990. Although the CBD and all rings experienced significant growth in income during the two decades, it is useful to compare the CBD and ring growth rates with the total region growth rate. This rate was slightly over 100% for both the 1970s and the 1980s. The CBD's growth was slightly under or about equal to the region's growth rate since the 1970s, while the core counties' growth rate was less (significantly less during the 1970s). The suburbs grew as income earners, however. The inner suburbs' growth rates were slightly over 100% during both decades,

TABLE 2.2 Earnings by Place of Work for the New York Metropolitan Region, 1970-1990

	Earnings by Place of Work (in millions of dollars)			Percentage Change		Percentage of Region Total		
	1970	1980	1990	1970-1980	1980-1990	1970	1980	1990
Regional total	73,371	153,413	330,647	109.1	115.5			
CBD	27,769	53,767	116,717	93.6	117.0	37.8	35.0	35.3
Core counties	13,690	23,699	44,860	73.1	89.3	18.7	15.4	13.6
Inner suburbs	19,529	41,964	85,041	114.9	102.7	26.6	27.4	25.7
Outer suburbs	12,381	33,981	84,027	174.5	147.3	16.9	22.2	25.4

SOURCE: Data taken from Regional Economic Information System, Bureau of Economic Analysis, U.S. Department of Commerce.

and the outer suburbs' growth rates were well over 100% during both decades.

Growth rates are not necessarily reliable indicators of economic importance of the different parts of the region, because the various segments are all growing from economic bases of different sizes. We should underscore these as differences in relative growth. Suburban growth takes place on a slender base, and modest increases are translated into large percentages. Obversely, the CBD and core counties have much larger bases, and percentage increases are translated as modest rates of growth.

Nevertheless, there are changes in the economic dynamics of the region, and these can be gauged by looking at the share of total regional income earned by place of work during the two decades. The CBD remained the dominant regional subdivision in terms of producing income earned by place of work, yet when it is combined with the core counties' share of the region's earnings by place of work, the central core's share drops from 56.5% of the region total in 1970 to 48.9% of the region total in 1990. In contrast, the outer suburbs' share of total regional income earned by place of work increased nearly 9% during the two decades, the only part of the region to experience an increase.

In examining the formation of new suburban agglomeration economies, Stanback suggests that the growth of metropolitan outer "cities" can be measured by looking at two indicators (Stanback, 1991, p. 65). One is the employment-population ratio of jurisdictions. High ratios (and increases in the ratios) of employment to population are indicative of substantial agglomeration because they signal areas where employment is becoming more important relative to residence. The other indicator is

TABLE 2.3 Employment/Population Ratio for the New York Metropolitan Region, 1970-1990

	1970	1980	1990
Total region	.47	.50	.56
CBD	1.80	1.72	1.76
Core counties	.25	.25	.27
Inner suburbs	.45	.52	.60
Outer suburbs	.36	.44	.56

SOURCE: Data taken from Regional Economic Information System, Bureau of Economic Analysis, U.S. Department of Commerce.

shift in the residence adjustment earnings, expressed as a percentage of the total earnings of the county workforce.

Residence adjustment measures net commuter income. When the amount is positive, it means that out-commuter income is greater than in-commuter income. That is, those residents of a given jurisdiction (out commuters) who work outside the jurisdiction bring in more commuter income than residents from other jurisdictions take out. When the amount is negative, it means that in-commuter income exceeds out-commuter income. That is, the jurisdiction's own residents are bringing in less income from their work in other jurisdictions than residents from other jurisdictions (in-commuters) are taking out. Residence adjustments focus on commuters and say nothing about the total amount of income being earned within a given jurisdiction.

The employment/population ratios for the region, the CBD, and its rings shows that the region's ratio of jobs to population has increased only slightly over the last two decades (Table 2.3). The CBD has maintained its dominant position, although it has declined slightly. The core counties of New York's outer boroughs and Hudson County have the lowest ratios. The inner and outer suburbs dramatically increased their ratios of jobs to population, with the outer suburbs showing the greatest growth. Although the outer suburbs were well below the aggregate regional ratio in 1970, they became almost equal to the regional ratio in 1990.

Resident adjustment data suggest a continued monetary flow outward from the CBD as well as increased amounts coming into core and suburban counties from other areas. Table 2.4 displays these movements by commuters since 1970. The table also displays the residence adjustment as a percentage of total income earned by place of residence. It shows that the CBD exported more income as a result of commutation than its

TABLE 2.4 Residence Adjustment for the New York Metropolitan Region, 1970-1990

	Residence Adjustment (millions of dollars)			Residence Adjustment as a Percentage of Income Earned by Place of Residence		
	1970	1980	1990	1970	1980	1990
Region total	−373	−2,226	−48,901	−0.53	−0.96	−16.2
CBD	−17,796	−33,182	−64,938	−200	−188	−150
Core counties	9,364	13,865	31,738	41.7	38.4	43.4
Inner suburbs	3,367	5,435	8,947	15.2	12.0	10.1
Outer suburbs	4,691	11,656	19,363	28.3	26.6	19.8

SOURCE: Data taken from Regional Economic Information System, Bureau of Economic Analysis, U.S. Department of Commerce.

residents earned during the two decades. The three surrounding rings all have positive percentages, so that other counties generated income for residents. Major changes, however, have been occurring in the suburbs. Both the inner suburbs' and outer suburbs' percentages declined over the two decades, suggesting a vital growth of employment within these areas.

Close examination of key types of employment suggest that the suburbs are growing at a much more rapid rate than the CBD and capturing new economic functions in "edge cities." The rise of new suburban agglomeration economies may be occurring at the expense of the CBD and core counties. The Manhattan CBD traditionally has played the leading role in certain producer and service industries (Drennan, 1989; Port Authority of New York and New Jersey, 1994a). Among these are employment in the finance, insurance, and real estate (FIRE) sector, the service sector, and wholesaling. In relative terms, the CBD's share of employment in these key sectors has declined over the past two decades.

As one of the growth sectors of the postindustrial economy, the FIRE industries have become a major new component of the region's economic health. As Table 2.5 shows, the region overall experienced growth in this sector during both decades, especially in the 1980s. Despite the CBD's slight decline in its percentage of total regional employment in the FIRE sector over the two decades, it clearly remains the dominant regional subdivision in this sector. FIRE employment grew as a component of the CBD's economy over the two decades being examined. Like the CBD, the core counties lost FIRE jobs in the 1970s, but they did not recover in the 1980s.

TABLE 2.5 Employment in Selected Sectors for the New York Metropolitan Region, 1970-1990

	Percent Change		Percent of Ring/CBD Total Employment			Percent of Region Sector Employment		
	1970-1980	*1980-1990*	*1970*	*1980*	*1990*	*1970*	*1980*	*1990*
Finance, Insurance, and Real Estate (FIRE)								
Region Total	7.0	22.7	10.1	10.4	11.1			
CBD	−1.1	13.9	15.7	17.4	18.5	50.3	46.6	43.2
Core counties	−14.9	3.0	7.4	7.1	6.6	15.7	12.4	10.4
Inner suburbs	10.8	20.9	8.2	8.4	9.1	22.5	23.3	22.8
Outer suburbs	66.0	62.5	6.1	7.4	9.1	11.5	17.7	23.4
Service								
Region total	26.3	41.4	21.6	26.4	32.4			
CBD	5.6	23.6	25.3	30.1	34.8	38.1	31.9	27.9
Core counties	24.8	37.1	20.0	28.0	34.7	19.6	19.3	18.7
Inner suburbs	32.5	42.5	20.6	25.3	32.4	26.4	27.7	27.9
Outer suburbs	67.9	70.9	18.2	22.3	28.8	15.8	21.0	25.3
Wholesale								
Region total	13.7	10.8	6.3	7.0	6.7			
CBD	−19.6	−18.6	8.5	7.7	5.8	43.5	30.7	22.5
Core counties	3.1	13.7	5.7	6.6	6.8	19.1	17.3	17.8
Inner suburbs	38.8	11.2	6.2	8.0	8.0	27.1	33.1	33.2
Outer suburbs	110.6	56.0	3.4	5.2	6.2	10.1	18.6	26.3

SOURCE: Data taken from Regional Economic Information System, Bureau of Economic Analysis, U.S. Department of Commerce.

FIRE employment has helped fuel the booming economy of the suburbs. FIRE employment in outer suburbs grew dramatically during the two decades (66.0% and 62.5%). Although FIRE employment in the outer suburbs experienced little growth as a percentage of their own economy, the outer suburbs' FIRE employment as a percentage of regional FIRE employment actually doubled, underscoring the growing role of the outer suburbs in the region's economy and, at the same time, establishing the outer suburbs as a major employment center in this sector. FIRE employment in the inner suburbs also has grown since 1970, although the regional share and the share of FIRE employment as a component of the inner suburbs' economy did not change dramatically. Together, however, the inner and outer suburbs' share of the region's FIRE employment increased from 33.8% to 46.2%.

In the service and wholesale employment sectors, the CBD percentages of all regional employment declined while that of the core counties and the inner suburbs remained relatively stable. In contrast, the outer suburbs' percentage of regional service and wholesale employment grew at a much more rapid rate than in any of the other rings. Over the two decades being examined, the combined suburban share of regional service and wholesale employment increased from well below to well above the majority of jobs in the region for each of these job sectors, while the CBD and inner core fell to below half. Thus, not only is employment in the suburbs, and especially the outer suburbs, growing at a faster rate than in the CBD, but the suburbs also now hold a majority of the region's job base in several key employment sectors that previously were located in the central city areas.

■ The Postindustrial Region and Governmental Cooperation

In effect, postindustrial economic changes have been forging new economic relationships within the New York metropolitan region. Although Manhattan still retains its position as the center of regional economic activity, the New York region has become increasingly polycentric. Other counties now compete in key economic subsectors, such as FIRE, where the CBD once held a monopoly. Although CBD employment growth rates in various sectors of the economy are positive, the CBD is no longer the singular, dominant economic force. The change in the CBD's economic status remains a sensitive issue for its political elites, who have, in many cases, adopted a competitive posture toward the movement of businesses and jobs in and out of the CBD.

This shift in the CBD's status as the region's economic center probably would not be perceived as being as serious a condition if the CBD were not appended to most of the core counties (except for Hudson County, New Jersey). Of the rings surrounding the CBD, the core counties are in the weakest economic position. In 1970, their economic position was weak, and it has declined since then. Their employment growth rates are the lowest in the region, and in many cases they are negative. They are the most dependent on the other regional subdivisions for employment of their residents. This type of uneven economic development increasingly has been the cause of political friction between Manhattan and its outer borough counties (Mollenkopf & Castells, 1991; Mollenkopf, 1992). Indeed, it is at the root of a recent movement for Staten Island's secession from New York City (Brown, 1994).

Stanback's concept of suburban agglomeration appears to be an accurate description of the postindustrial economic changes that have affected the inner and outer suburbs. The impacts of the economic shifts on the two suburban rings, however, are quite different. The inner suburbs are the most stable of the region's subdivisions. Over the two decades, this segment of the region has grown slowly, but it has not declined. These suburbs are acquiring many of the characteristics of a maturing suburban economy.

The outer suburbs are a different story. They experienced the greatest growth since 1970. Although some outer suburbs in 1970 were in a weaker economic position than were the core counties, by 1990 communities in this ring had become leading participants in new economic sectors. New competing economic interests within the region are evolving. In fact, there may be greater hostility toward regional cooperation if it means sacrificing some of the newly gained economic strength of suburban communities.

■ Political Entrepreneurship and Regional Cooperation

The region's political institutions and players often reacted to the new forces of urban change by voicing support for more cooperative solutions during the last several decades. Greater attention to regional awareness among political entrepreneurs in the cities, suburbs, and states, however, very rarely translated into actual political cooperation.

The Decline of Regional Forums of Cooperation

Lack of a regional government in such a crowded and interdependent economic zone has long prompted interest in forums for intergovernmental cooperation, yet the tristate area has not been able to sustain a confederational council with any power over its members. In the private sector, there has been some interest in promoting regional planning through the Regional Plan Association. The Regional Plan Association (RPA), in existence since 1929, is a private nonprofit group that serves as the metropolitan area's only consistent advocate for regional cooperation. Although it does receive some funding from restricted state and local government grants, the RPA receives no earmarked public funding and has no elected officials or representatives of state or local governments on its board of directors. It relies heavily on corporate support. The RPA is limited by its lack of official status or recognition. In recent years, its primary activity has been environmental and infrastructure research (Lueck, 1993).

Attempts to use confederational approaches to create intergovernmental cooperation have foundered. Because confederational councils have no powers other than those given to them by their members, support for them is vulnerable as the number of member governments grows and interests diverge. The tristate character and large number of local governments in the New York region have always undermined this form of intergovernmental coordination.

The life and death of the Tri-State Regional Planning Commission illustrates these realities. During the 1960s, there were several attempts to get the region's governments to pull together. One attempt, the New York Metropolitan Regional Council, failed to gain any official status or operating authority because there was always some group of governments in the region unwilling to support the council's plans (Aron, 1969, p. 25). After federal authorities threatened to cut off highway and other aid to the region unless a formal planning agency was started, an informal transportation planning committee was organized. In 1971, it was named the Tri-State Regional Planning Commission.

The commission foundered from the beginning because of its lack of legitimacy and authority. Its members were state representatives rather than representatives of local governments in the region. Tri-State clearly was more of an interstate regional commission than an interlocal one. In addition, the commission was deprived of all authority except planning functions at the request of the New Jersey state legislature (Danielson & Doig, 1982, p. 168; "Jersey Advances," 1965). By the end of the 1970s, Tri-State's activities had become largely unnoticed, yet it was heavily criticized for delays in obtaining federal funds, wasting tax dollars, and rubber-stamping approval of grant applications from local agencies, however poorly conceived or costly they were (McQuinton, 1979). Governmental support for Tri-State dwindled rapidly as all three states acted to narrow the commission's review of programs (Danielson & Doig, 1982, p. 170). After the Reagan Administration decreased federal funding to support regional councils, the bulk of Tri-State's financial support evaporated; because the state governments were unwilling to make up for federal cuts, Tri-State died.

■ The Only Planning Game in Town: The Port Authority and Regional Politics

As the only public regional planning agency, the Port Authority of New York and New Jersey is in a pivotal position to promote greater regional cooperation. With its staff of more than 9,200, gross operating revenues of $1.9 billion, and 34 projects scattered over the authority's 17-county region (Port Authority of New York and New Jersey, 1994a, p. 2), the Port Authority (PA) is, in effect, the "only planning game in town" (Walsh & Leigland, 1983). Indeed, the PA claims that its own activities support more than 400,000 jobs in the region (Port Authority of New York and New Jersey, 1994b, p. 88). Its enormous revenues and substantial surpluses from a number of money-making enterprises, such as toll bridges, tunnels, and airports, give the PA great leverage as a regional planner and make it a tempting target for politicians wishing to harness its revenues.

It is difficult, nevertheless, for the PA to function as a regional planner. Prior to 1970, the PA had shown little disposition to take up this role. When the bistate agency was created in 1921, it failed in its first job of getting rail interests to agree to a beltway system that could facilitate the movement of freight to and from the port. After the railroads refused to cooperate and the authority lost other battles during its early years, the PA turned to easier ways of winning power (Doig, 1993; Moss, 1988). The PA settled into building and operating low-risk and high-return infrastructure projects. This strategy eventually made it the nation's richest and most powerful regional agency.

PA activities deviated from this strategy only when the governors of New York and New Jersey, supported by their legislatures, intruded to force changes in direction. For example, during the 1960s, the PA was forced to assume some responsibility for mass transit (it took over a failing New Jersey railroad). Even this change required a protracted political struggle by the bistate governors, who made the authority's intransigence a major public issue.

History aside, the governing structure of the PA does not easily support a leadership role in building regional cooperation. It is run by 12 commissioners, 6 each appointed by the governors of New York and New Jersey. Although the governors can veto PA actions and their pressure can be brought to bear on the authority, the commissioners financially are quite independent. The authority does not levy taxes or depend on subsidies from other governments; it derives its revenues from the projects it builds and operates. This financial independence, together with the authority's legal responsibilities to bondholders, gives the authority little reason to reach out to the region's other political interests by including them in the agency's planning activities.

The persistence of these obstacles to regional leadership by the PA are confirmed by events. In the 1970s, the PA began to recognize the need for more attention to regional planning of economic development. Although it made an attempt to respond to the need, the agency largely failed to bring about greater regional political cooperation or to assert its own power in order to compensate for noncooperation among the other governmental players in the region. Circumstances during the late 1970s moved the agency to entertain an enlarged role as a regional development planner. The long recession of the early 1970s, followed by New York's fiscal crisis, weakened the whole regional economy and awakened many politicians to the desirability of achieving economic revival through better regional cooperation and forecasting. During the 1980s, the governors of New York and New Jersey put pressure on the PA to assume a larger role in regional development. New Jersey governor Thomas Kean and New York's Hugh Cary appointed a bistate panel to look into the revenue generating capacity of the authority and the ways it could be used to provide more regional benefits. Later, the governors threatened to sell off one of the PA's most important "cash cows," the World Trade Center, which accounted for 22% of the PA's revenues in 1981 (Walsh & Leigland, 1983, p. 9). At the same time, a new cast of political entrepreneurs emerged. In 1977, Peter C. Goldmark, Jr., an advocate of greater PA leadership in regional economic development, became executive director of the agency. He organized a number of planning committees that dealt with the economic challenges facing the region, including a Committee on the Future that tried to develop a regional economic development plan in 1979.

In part to head off further gubernatorial threats, Goldmark and later directors adopted a bevy of regional development programs in response to deals struck by the bistate governors. They included such projects as a teleport (satellite communications park) in Staten Island, office buildings, a resource recovery plant, waterfront projects, and industrial parks as well as a development bank to repair public works (Roberts, 1984). The PA also became an advocate of regional planning and cooperation. It set up a study of regional economic competition and issued plans that outlined regional investment and marketing strategies (Roberts, 1989). It also organized forums on regional cooperation that led to a 1991 pact among New York City and the governors of the three states to cease pirating businesses from one another.

Despite this shift in agenda, it is doubtful if policy changes have significantly increased political cooperation on matters of economic development. The regional development programs never became a very large part of the authority's activities. The operating revenues of all the regional development programs amount to only $93 million out of more than $1.9 billion in total PA operating revenues, and the programs impose an operating loss on the PA's books of merely $8 million (Port Authority of New York and New Jersey, 1994a, p. 80).

The PA's new regional development programs never found stable political support within the region. Because the economic development projects often met with resistance within the agency, it usually took prodding and threats by the governors to get the agency to move ahead with them (Walsh & Leigland, 1983, p. 209). For example, a legal center in Newark was built only after New Jersey governor Kean told the agency that it would have to commit itself to the project if it wanted his support for a toll increase in 1983 (Finder, 1991). Furthermore, New York City officials opposed enlarging the PA's development role. New York's mayor, Ed Koch, urged sale of the World Trade Center and opposed a regional development bank program (Lueck, 1988). He contended that authority ownership of the twin towers deprived the city of as much as $100 million a year in real estate taxes and that a regional program would subsidize construction projects in competing New Jersey ("City Against," May 15, 1984: A3).

Since 1985, less emphasis has been placed on regional development by PA directors. The economic doldrums of the 1990s have left PA revenues flat, and its current leaders believe that it must meet more fundamental needs, particularly improving the region's airports and other infrastructure. At the same time, the failure of some projects has made the agency wary of development ventures. For example, a $25 million fish market in South Brooklyn never attracted boats or wholesalers (Finder, 1991).

Promotion of regional cooperation by the Port Authority also has remained elusive. First, the agency never developed a mechanism that permitted participation by representatives of cities or communities in developing PA projects. With its money-generating orientation, the authority has been unable to supply the political entrepreneurship necessary for building on regionwide interests. In 1979 and 1986, PA task forces proposed regional strategies for investment and regeneration (Committee on the Future, Port Authority of New York and New Jersey, 1979; Port Authority of New York and New Jersey, 1986). Although these proposals could have formed the basis for a comprehensive regional economic plan (Rubin & Wagner, 1981), the PA made no attempt to specify how regional political agreement could be forged for strategy implementation, other than to say that coalition building was necessary (Port Authority of New York and New Jersey, 1986, p. iv).

Second, the PA's appeals for regional cooperation have been insufficient to overcome the intense economic rivalry among political jurisdictions in the tristate area. In 1991, the executives of the three states and New York City signed a "nonaggression" pact that was quite limited in its promises. The four governments vowed not to use negative advertising against one another and to avoid business incentive programs designed to attract companies within the region. They also agreed to increase cooperation in marketing, develop a regional development strategy, and create a committee of policymakers to meet periodically in order to iron out common problems and plan regional initiatives. Although there was some diminution of negative advertising and minor forms of cooperation, within less than a year, rivalry among all the governments over luring jobs from one another undermined any semblance of upholding the pact. All three states launched new business incentive and tax reduction programs. Job wars between New York City and New Jersey accounted for the largest breaches in the pact (Salamore & Salamore, 1993). For example, New Jersey induced the First Chicago Trust Company to transfer 1,000 jobs from lower Manhattan through a program that offered the company a subsidized office for moving to the state. New York State's economic development commissioner threatened to retaliate by starting a program specifically to lure New Jersey firms to move across the Hudson. Ironically, the source of funding for New Jersey's $224 million program (that New York officials claimed was stealing the jobs) was bonds that essentially will be repaid by New Jersey's share of rental income from the Port Authority's World Trade Center (Peterson, 1992; Prokesch, 1992a, 1992b).

The Politics of Limited Cooperation: Mass Transit

Although regionwide political cooperation has not been very successful, there have been some instances of more limited intergovernmental
coordination. These relatively successful attempts to overcome political
fragmentation have occurred when the number of players in the arena has
been limited and when the issues have been confined to a single policy
area.

The response of New York's Metropolitan Transit Authority (MTA) to
the transit crisis that enveloped the region during the 1980s is such a case.
The giant MTA was created during the 1960s to integrate mass transit in
12 counties of the New York part of the region. The agency assumed
control of the New York City subway and public bus systems, which were
folded together with the commuter railroads operated by the Long Island
Railroad, the New York Central Railroad, and the New Haven Railroad to
create a system of transit authorities under a single management. MTA
operations are critical to the region as a whole, particularly because of the
supreme economic importance of mass transit for the Manhattan CBD
(Metropolitan Transit Authority, n.d.; Seeley, 1982).

Unlike the Port Authority, the MTA always has required substantial
state and regional political cooperation in order to function. For one thing,
the political structure of the MTA devolves some power to governments
in the region. Although the MTA board of 17 members is appointed by the
governor with advice and consent of the New York State Senate, 11 of the
appointments are made on the recommendation of local officials. Four are
nominated by the mayor of New York City, one each is nominated by the
county executives of Westchester, Nassau, and Suffolk counties, and one
is nominated by the county executives of Rockland, Orange, Putnam, and
Duchess counties. These 17 members have only 14 votes, however, because the suburban county representatives have only a quarter vote each.

Political cooperation among city, state, and suburban interests is necessary because all these jurisdictions provide subsidies for capital and
operating expenses of the MTA authorities. As in most mass transit
operations, fares are unable to cover expenses by a wide margin, and the
huge, aging transit system has a voracious need for continuing capital
improvements.

The MTA also depends on considerable state-level cooperation. Not
only does it lack money-generating projects, but mass transit does not
mobilize a powerful and unified voter constituency (Danielson & Doig,

1982, p. 213). Dependent on state fiscal largesse, the MTA is compelled to seek the protection of powerful political patrons in Albany. When the agency was formed under Nelson Rockefeller, a powerful governor, the agency was run with little interference from others (Walsh, 1978, p. 274). After Rockefeller departed, however, the agency remained adrift and dependent on the shifting tides in the annual legislative appropriations process (in the state and in New York City) and New York State general obligation bonds for its financing.

The problems emanating from the MTA's fiscal dependency and weak leadership were exacerbated by the fiscal crisis that hit both state and local governments during the 1970s. The weak condition of the state and city economies led to cutbacks and fare hikes that failed to keep up with the transit system's capital needs. Years of deferred maintenance and obsolete equipment left a decayed and dangerous subway system that was losing riders rapidly. In 1980, the state legislature declared a transportation emergency.

MTA head Richard Ravitch capitalized on the crisis to put together an ambitious capital program. Ravitch won backing in Albany for changes that enabled the transportation authority to undertake a huge ($16 billion) rolling capital program for more than ten years (Metropolitan Transit Authority, 1990, p. 3). Most important, Ravitch also managed to diminish significantly the MTA's dependence on annual legislative funding and bonding that had crippled long-range program planning in the past. Legislation was passed in 1981 that provided the MTA with a number of new revenue streams that did not depend on annual authorization, including new state taxes levied with the revenue dedicated to MTA programs. A capital plan review board composed of representatives of the two houses of the legislature, New York City, and the suburbs also was set up to monitor the MTA programs. At the same time, Ravitch deflected a move by Governor Mario Cuomo to bring the MTA under his direct control.

Although these changes did not free the MTA from coping with deficits, the political engineering of Ravitch and his successor, Richard Kiley, permitted the MTA to undertake an unprecedented regional reinvestment program without creating conflicts among key interests. In particular, such a huge public works program could have pitted city against suburb. The MTA board is organized to give proportionate weight to New York City, state, and suburban claims. The state review board does the same thing through its membership; the Republican-controlled Senate traditionally favors the suburbs, while the Democrat-dominated Assembly is more sensitive to New York City.

Conflicts within the MTA, its review board, and its legislative over-seers were limited because these participants agreed not to contest the division of capital monies. Instead, they agreed to divide new capital monies in a ratio of 77% for the New York City Transit and 23% for the commuter railroads that serve the suburbs. This distributive formula has remained unchallenged since it was agreed upon during the 1970s, after the MTA was created. Participants recognize that there is more to gain by maintaining a united front in order to limit legislative interference and maximize overall state funding rather than by fighting over issues of distribution. Further, suburban representatives tend to regard 23% for the suburban rail lines as generous in a system in which New York City's transit woes are invariably expensive (Peter Derrick, personal communication, July 13, 1994).

During the 1990s, transit planning remains relatively consensual. Having established its credibility as a capital planner during the 1980s, and owning a more stable financial base, the MTA members are inclined to close ranks in order to satisfy capital needs. Conflicts between city and suburban interests are emerging, however. The MTA's current chairman, Peter E. Strangl, lacks many of the political skills of his predecessors, and the region's flat economy is triggering greater competition for resources and savings. Regional conflicts that percolated up to the state legislature delayed the authority's third 5-year capital plan for nearly 6 months during 1993 and 1994. This time, Albany lawmakers thrashed out details of how some of the money should be spent and remained deadlocked over one suburban rail project that represented less than 0.1% of the entire plan. At the same time, New York City officials balked over providing $500 million in additional funds to the MTA for bus and subway improvements (Dao, 1993). Nevertheless, transportation planning remains one of the more cooperative endeavors in the region.

■ The Consequences of Noncooperation: Policy Voids and Inequality

Given the limited cooperation that characterizes the region, some of the most serious social and economic challenges of postindustrialism have fallen into policy voids. These problems become part of the agendas of the lowest levels of government in the region, where officials compete to avoid confronting them or seek to shift their social costs onto other governmental jurisdictions. This form of political competition increases social inequalities.

The economic competition for jobs among states and localities in the tristate region goes unregulated, to the disadvantage of almost everyone except the corporations that are objects of subsidies. The subsidized office parks, tax abatements, and other lures that area governments use in hopes of gaining jobs from one another is a zero sum game that depletes the availability of tax revenues for more productive uses, such as better schools or roads. Although this competition could take the form of using funds to improve human capital (for education or job training, for example) and to invest in physical infrastructure, in practice local governments are inclined to compete through programs to lower business costs, not by increasing business productivity (Kantor, 1995, chap. 5). The highly fragmented character of the New York metropolitan region accentuates this tendency. In 1991, New York City lost about $700 million in the form of tax abatements—five times the yearly budget of the city's library system. The cost of such programs to New Jersey and Connecticut probably is similar (Reich, 1988, p. 34).

The greatest policy voids occur in the area of housing, because the local governments have so much power over these programs and because competition among governments to keep out the "undesirables"—meaning poor people, minorities, homeless people, and others perceived to cause central city-type problems—is seen as a key to the good suburban lifestyle. The result not only is that the New York region is characterized by great inequality in housing opportunities and schooling among communities but also is that lack of regional cooperation has made reform efforts almost wholly dependent on court suits and state government intervention (Hanley, 1987). In effect, the region's limited capability for political cooperation over matters of social equality has displaced these matters to outside authorities by default.

Only when local governments of the region are more dependent on state fiscal assistance have court mandates had greater effect. For example, in education, court judgments have prodded state governments to adjust school aid formulas to bring about greater parity in school financing among local school districts. During the last two decades, court suits in all three states have forced state legislatures to alter aid formulas, with the result that there is much greater equality in spending among school districts in the region than ever before. For example, in New Jersey, poor districts now spend about 84% of what wealthy districts spend. In recent court cases, questions of equalized spending have grown to include the effectiveness of education in poor districts, leading to mandates for compensatory programs in needy areas ("Top New Jersey Court," 1994, A1, B6).

The Accelerating Politics
of Conflict and Competition

The process of regional political development seems likely to render greater political cooperation more difficult to achieve in the future. Changes in regional employment, wealth, and economic activities in recent decades are generating the formation of new political interests within the region. These competing interests are likely to further frustrate political cooperation between central city and suburb and even within suburban areas, accelerating the region's political tradition of fragmentation and rivalry.

Manhattan's traditional CBD functions remain, but some of them increasingly are being spun off to suburbia as agglomeration economies develop in the periphery. In addition, a "dual suburbia" is now emerging. It is no longer accurate to speak of city versus suburban interests in the regional economy. Older suburbs are maturing into steady state or slow-growth economies while newer exurban communities experience rapid growth and are capturing new economic functions that compete with other areas.

The region's political entrepreneurs generally have not responded by forging many cooperative solutions to the region's new problems. Political institutions for building broad metropolitan alliances have yet to emerge. Older institutions having wide functional scope are ill suited to the task, or they lack the power and support for such ventures. At best, efforts to meet common challenges have been successful only when the number of players and the issues are limited in scope. Without more creative regional cooperation, problem solving will continue to fall by default into the hands of the courts, the states, or the federal government.

NOTES

1. There is no standard definition of the New York region. For a discussion of different regional concepts, see Danielson and Doig (1982, Chapter 2) and Harris (1991).

2. In 1898, the government of Greater New York was created by consolidating New York and its surrounding four counties of Kings (Brooklyn), Richmond (Staten Island), Queens, and the Bronx. At the turn of the century, this was a government of almost regional scope.

3. Ring 2 inner counties comprise the northern New Jersey counties of Bergen, Essex, Passaic, and Union and two New York counties, Westchester and Nassau. The outer counties of Ring 3 include Fairfield in Connecticut; Rockland, Orange, Putnam, and Suffolk in New York; and Hunterdon, Middlesex, Monmouth, Morris, Ocean, Somerset and Sussex in New Jersey. The New York region is not exactly coterminous with the CMSA because the latter includes only parts of some of the counties. For a somewhat similar approach to the region, see Harris (1991).

4. Postindustrial change also has brought about social restructuring, including changes in immigration, education, social networks, family structures and lifestyles (Mollenkopf & Castells, 1991; Sassen, 1991). These changes are not within the scope of this analysis.

REFERENCES

Aron, J. (1969). *The quest for regional cooperation.* Berkeley: University of California Press.

Beauregard, R. A. (1991). Capital restructuring and the new built environment. *International Journal of Urban and Regional Research, 15,* 90-105.

Brown, C. (1994, January 30). Escape from New York. *New York Times Magazine,* pp. 22-26, 43-59.

Bureau of the Census. (1992). *Census of governments.* Washington, DC: Government Printing Office.

City against regional fund unless trade center is sold. (1984, May 15). *New York Times,* p. A3.

Committee on the Future, Port Authority of New York and New Jersey. (1979). *Regional and economic development strategies for the 1980s.* New York: Author.

Danielson, M. N., & Doig, J. W. (1982). *New York: The politics of urban regional development.* Berkeley: University of California Press.

Dao, J. (1993, December 18). Mass transit gets millions from Albany. *New York Times,* p. B1.

Doig, J. W. (1993). *The 1920s as a regional battlefield: New institutions and the search for efficiency.* Unpublished manuscript, Princeton University.

Drennan, M. P. (1989). The local economy. In Charles B. Horton & R. D. Horton (Eds.), *Setting municipal priorities* (pp. 27-49). New York: New York University Press.

Finder, A. (1991, November 22). In the eye of the storm, the Port Authority battens down. *New York Times,* pp. B1, B2.

Hanley, R. (1987, June 1). Housing the poor in suburbia: A failed vision in New Jersey. *New York Times,* p. B1, B4.

Harris, R. (1991). The geography of employment and residence in New York since 1950. In J. Mollenkopf & M. Castells (Eds.), *Dual city* (pp. 129-152). New York: Russell Sage.

Hoover, E. M., & Vernon, R. (1959). *Anatomy of a metropolis.* Cambridge, MA: Harvard University Press.

Jersey advances 3-state rail plan. (1965, March 25). *New York Times,* p. B3.

Kantor, P. (1995). *The dependent city revisited: The political economy of urban economic and social policy.* Boulder, CO: Westview.

Lloyd, E., & Kahn, K. (1991). Bringing global business to the region. *Portfolio, 4*(Autumn), 37-44.

Lueck, T. (1988, February 26). Study sees threats to New York region's growth. *New York Times,* p. B3.

Lueck, T. (1993, June 8). Protection of regional open spaces urged by planning group. *New York Times,* p. B1.

McQuinton, J. (1979, February 4). Panel to examine tristate commission. *New York Times,* p. H9.

Metropolitan Transit Authority. (n.d.). *No standing still: The MTA capital program, Phase 3, 1992-1996.* New York: Author.

Metropolitan Transit Authority. (1990). *MTA capital needs and opportunities, 1992-2011.* New York: Author.

Mollenkopf, J. (1992) *Phoenix in the ashes.* Princeton, NJ: Princeton University Press.

Mollenkopf, J., & Castells, M. (Eds.). (1991). *Dual city: Restructuring New York.* New York: Russell Sage.

Moss, M. L. (1988). New York vs. New Jersey: A new perspective. *Portfolio, 1*(Summer), 23-30.

Peterson, I. (1992, October 7). New cash subsidies help Trenton lure businesses. *New York Times,* p. B6.

Port Authority of New York and New Jersey. (1986). *Regional competition panel, progress report 1985.* New York: Author.

Port Authority of New York and New Jersey. (1994a). *Comprehensive annual financial report, 1993.* New York: Author.

Port Authority of New York and New Jersey. (1994b). *Regional economy: Review 1993, outlook 1994 for the New York regional economy.* New York: Author.

Prokesch, S. (1992a, November 30). Promises aside, states in region fight with one another for jobs. *New York Times,* pp. A1, B2.

Prokesch, S. (1992b, December 1). New Jersey and New York collide in new competition to lure jobs. *New York Times,* pp. A1, B6.

Reich, R. B. (1988). Making America more competitive. *Portfolio, 1*(Summer), 3-15.

Roberts, S. (1984, September 9). Goldmark says he will leave the Port Authority. *New York Times,* p. B1.

Roberts, S. (1989, February 2). Spanish lesson: The solutions are regional. *New York Times,* p. B1.

Rubin, M., & Wagner, I. (1981). A regional economic development strategy. In B. J. Klebaner (Ed.), *New York's changing economic base* (pp. 124-131). New York: Pica.

Salamore, B. G., & Salamore, S. A. (1993). *New Jersey politics and government.* Lincoln: University of Nebraska Press.

Sanders, H., & Stone, C. (Eds.). (1987). *The politics of urban development.* Lawrence: University of Kansas Press.

Sassen, S. (1991). *The global city.* Princeton, NJ: Princeton University Press.

Seeley, E. S., Jr. (1982). Mass transit. In C. Brecher & R. D. Horton (Eds.), *Setting municipal priorities, 1982* (pp. 394-415). New York: Russell Sage.

Stanback, T. M. (1991). *The new suburbanization.* Boulder, CO: Westview.

Stone, C. (1989). *Regime politics.* Lawrence: University of Kansas Press.

Top New Jersey court orders new plan for schools. (1994, July 13). *New York Times,* pp. A1, B6.

Walsh, A. (1978). *The public's business.* Cambridge, MA: MIT Press.

Walsh, A. (1991). Public authorities and the shape of decision making. In J. Bellush &
    D. Netzer (Eds.), *Urban politics, New York style* (pp. 188-222). Armonk, NY: M. E.
    Sharpe.
Walsh, A., & Leigland, J. (1983, May). The only planning game in town. *Empire State Report,*
    pp. 206-212.
Wood, R. (1961). *1400 governments.* Cambridge, MA: Harvard University Press.

3

Los Angeles: Politics Without Governance

ALAN L. SALTZSTEIN

Greater Los Angeles is a five-county region containing a population of more than 14 million, a land area nearly as large as Ohio, and a gross regional product that, were the region an independent country, would rank 12th largest in the world. The people of the region are very diverse racially and ethnically, and this diversity has increased markedly in the last 10 years. Sources of wealth and influence are unevenly dispersed throughout the urban area. The economy, once robust, remains in the midst of the most serious postwar recession and has undergone significant structural change. Consequently, differences of wealth have increased, and unemployment remains well above the national average. The area contains numerous governments with independent resources. Together, these conditions suggest that in spite of significant regional problems, developing strong regional government should be very difficult.

Regional governments, however, gained power and influence in greater Los Angeles during the past 10 years. A major program to control air pollution through intergovernmental cooperation was implemented. Citizens approved taxes for regional transportation, and mechanisms to coordinate transportation planning were developed. Preliminary approval was obtained for a regional comprehensive plan. Intervention by state and federal agencies, along with leadership of several elected officials, opened a brief "policy window" that led to the increased power of regional and subregional bodies.

As political and economic conditions changed, a negative reaction to regional governance followed. The future of regional government is now in doubt as local and subregional interests have exerted more influence on the policy-making process.

John W. Kingdon (1995) argues that changes in policy occur when streams of problems, policies, and politics come together at critical times (p. 194). "Policy windows," as he calls them, open briefly when ideas, political actors, and fortuitous times permit a "coupling" to occur. They remain open briefly and close as conditions change. Policy changes, then, move in fits and starts. As problems are perceived, solutions are selected from the "garbage can" of possible alternatives if conditions are present that encourage policy changes.

Kingdon's model seems to fit the recent history of regional politics in Los Angeles. Briefly, a combination of events encouraged regional policy making. Another series of events then closed the "window." Regionalism today is no longer an important component of the policy agenda, but regional institutions have more influence today than they did prior to the crystallization of the conditions that opened the window. Thus, following Kingdon, it is my argument that regional decisions followed from a confluence of economic conditions, leadership, and outside pressures. As conditions changed, the push for regional government also faded.

The steps toward regionalism in greater Los Angeles are influenced by major changes in the population, economy, and political system. The first section of this chapter examines these changes. The second provides a brief description of the political organization of greater Los Angeles, both regional and local. The third examines the emergence of regional actions in three areas—air pollution control, transportation, and regional planning —and the reaction to these policy changes as the conditions that led to their emergence as regional issues changed.

■ I. The Environment of Greater Los Angeles

Several characteristics of population and economy affect the ability of the region to make common decisions. One would assume that regions with a clearly defined core attached to outlying communities, a stable economy, and a population whose important characteristics do not change radically over time would be more likely to adopt regional solutions to public problems. Common experiences, commitments, and interests should facilitate the making of intergovernmental decisions. Successful regional agencies are found more frequently in urban areas with this kind of a profile.

None of these conditions are present in greater Los Angeles. The region is large in area, is increasingly dispersed, and has experienced immense changes in its population and economic base. "Within the last decade," state Lockwood and Leinberger, "greater Los Angeles has undergone economic, social, political, and cultural upheavals deeper and broader than those experienced by any large American city during the same period" (1988, p. 31). Thus, some of the difficulties in obtaining regional governance are a consequence of the social, demographic, and economic changes occurring in the region in the latter half of the 20th century.

The greater Los Angeles region comprises the five counties located in the Los Angeles basin: Ventura, Los Angeles, Orange, Riverside, and San Bernardino (Figure 2.1). With a few small exceptions, this area is the boundary of the major regional governments, the Southern California Association of Governments (SCAG), and the South Coast Air Quality Management District (AQMD). Most of the population of these counties is considered urbanized, and all residents are affected by a common set of metropolitan problems.

The City of Los Angeles is the logical center of the metropolis, with a population of 3.4 million, more than 20% of the total regional population. City boundaries, however, encompass a large area and contain many subcenters within them. Many sizable cities are dispersed throughout the region; 38 exceed 100,000 in population. Differences among the people of the region have increased in recent years. The population today is more widely spread, as well as more ethnically and racially heterogeneous, and centers of economic and commercial activity have expanded farther from downtown Los Angeles.

A. Demographics

Los Angeles has always been a decentralized urban area, and the City of Los Angeles, though unquestionably the central city, has never had the preeminence within the region held by other American cities. Population growth always has been shared between city and suburb in both Los Angeles and the adjacent counties. Since 1920, the suburban growth rate, in fact, always has been higher than that of the city (Soja, Morales, & Wolff, 1984, p. 207). Words like *city* and *suburb*, according to Lockwood and Leinberger (1988), are becoming meaningless in Los Angeles because they imply a metropolitan area that has followed the traditional development pattern, in which a dominant, highly residential, low-density

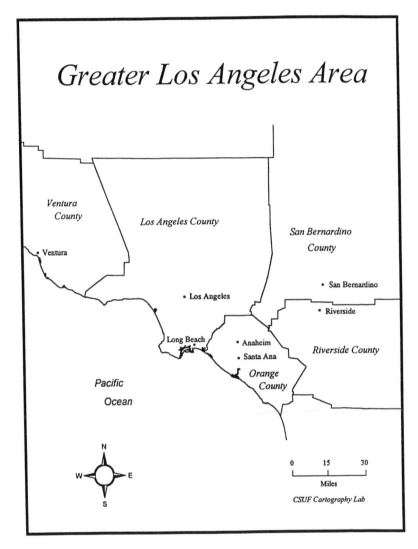

Greater Los Angeles Area

Figure 3.1.   The Greater Los Angeles Area

business center is surrounded by a dependent and mostly residential low-density periphery (p. 32).

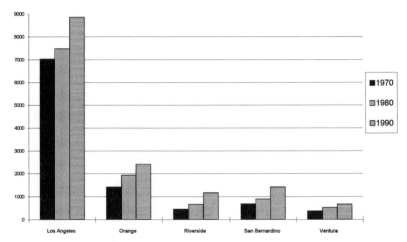

Figure 3.2. Population, Greater Los Angeles Counties, 1970-1990
(in thousands)

Recent years have witnessed increased growth in population in outlying areas (see Figure 3.2). From 1980 to 1990, the region as a whole grew
by more than 47%. The growth rates for the outlying counties far surpass
that of Los Angeles. Los Angeles County in 1990 comprised 60.5% of the
population of the region, down from 70.5% in 1970. The City of Los
Angeles also has become a smaller proportion of Los Angeles County,
declining from 37.6% of the county total in 1980 to 33.5% in 1990.

Historically, the distribution of racial and ethnic groups has followed
the traditional pattern, with concentrations of minority groups in the
central city. The 1990 census revealed a startling increase in minority
concentration within the City of Los Angeles. The Anglo proportion in
Los Angeles declined from 61.2% in 1980 to 37.3% in 1990. (The city
increased in population by 17% during this period). The 1990 census also
shows a major increase in diversity in the outlying areas. As Figure 3.3
indicates, the racial and ethnic minority proportion of the population
increased significantly in all counties from 1980 to 1990. All counties but
Ventura have seen minority increases of more than 10% in the past 10
years. The African American population, highly concentrated in South
Central Los Angeles in 1980, was more widely dispersed around the urban

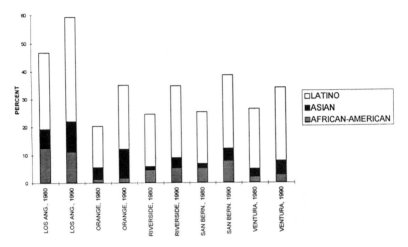

Figure 3.3.   The Changing Racial and Ethnic Population of Greater
Los Angeles, 1980-1990

core by 1990. Suburban areas within Los Angeles County and San Bernardino County added significant numbers of African Americans during the decade.

More than a million and a half Latinos were added to the urban area during the 1980s as a consequence of significant immigration from Mexico and Central America and a high birth rate. Recent immigrants were concentrated in traditional enclaves of East Los Angeles, Boyle Heights, and Pico-Union, and they displaced African Americans as the dominant group in South Central Los Angeles. Others were added throughout the region, with significant concentrations in central Orange County, the northern parts of the San Fernando Valley, and the cities of Riverside and San Bernardino ("The Next Los Angeles," 1994). Asian residents more than doubled during the 1980s and appeared to be randomly dispersed throughout the region, with the exception of South Central Los Angeles. The Asian population shows an immense variety of ethnicity, skills, and wealth. Geographic dispersion also differed; recent immigrants from Vietnam, for example, were concentrated in central Orange County; many Cambodians are found in Long Beach; and Koreans tended to live in central Los Angeles.

B. The Economy

The last two decades have been times of transition for the greater Los Angeles economy. Similar to national trends, manufacturing employment has experienced very little growth. More than 15,000 jobs in defense-related industries have been lost since the 1970s, much of that loss occurring in the last decade. Growth has occurred in professional services, tourism, and entertainment (Southern California Association of Governments, 1993, pp. 2-11).

The recession of the 1990s has had a telling effect on Southern California. The effects continued even while the economy of much of the rest of the nation improved. Unemployment in the region has risen since 1988. Nearly 700,000 people were unemployed in 1992, and more than half a million jobs were lost from 1990 to 1993 (Southern California Association of Governments, 1993, pp. 2-19).

There have been significant changes in the structure of employment as well. Jobs in the skilled trades are declining, with increases likely in clerical jobs and unskilled positions in service and tourism industries. Thus, middle-income jobs are being replaced by minimum wage positions, many of which are part-time. Job losses have been greater in central Los Angeles and in some of the older outlying areas. San Bernardino has been particularly affected by these changes. Its unemployment rate, at 11.3% in 1993, was the highest in the region.

Job gains, on the other hand, were more frequent in Orange County, where employment growth was projected to be the highest in California (Silverstein, 1988). Unemployment in Orange County was also much lower than that of the region as a whole.

Public revenues are directly related to the state of the economy. Revenue from sales taxes, property taxes, permits, and user fees have declined. Real taxable sales fell by 13.9% between 1989 and 1992, and they continue to fall (Southern California Association of Governments, 1993, pp. 2-22). The prices of existing homes have fallen since 1990. As state income has similarly declined, state aid to local governments has been curtailed. Significant property tax resources of both cities and counties were taken over by the state in 1993.

Poverty became more dispersed in the 1980s. Figure 3.4 compares the percentages of persons in poverty in the five counties from 1970 to 1990. A major increase in poverty occurred in San Bernardino County; more than one-fifth of its population was recorded as below the poverty level in 1990. The incidence of poverty also increased in Orange and Riverside Counties.

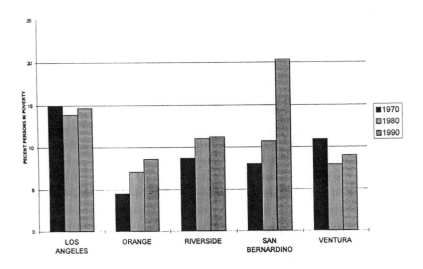

Figure 3.4. The Suburbanization of Poverty, 1970-1990

Together, these trends describe a region that has increased the potential sources of tension as it proceeds to make regional decisions. The population has increased in heterogeneity and spread further from the historic central city. Racial and ethnic differences have become greater, and these differences are present in all five counties. The economy has declined, and the presence of poverty is a trait of the newer areas as well as the traditional city center. Public revenues have been particularly hard hit by the problems of the economy.

■ II. Political Organization of Greater Los Angeles

Regional governance must coordinate existing governmental units. The greater Los Angeles area contains traditionally strong city and county governments. Most of the urbanized portion of the area is incorporated into one of more than 200 cities. These vary in size from Los Angeles, with a population in excess of 3 million, to several with populations less than 1,000. The larger communities are diverse cities with a cross section of residents and land uses. Among the small cities are some that specialize in a particular kind of land use and have very small populations.

All cities except Los Angeles and San Bernardino are governed by the council-manager form of government or a variant of it. Nonpartisanship and professional public management typify the area's political traditions. City councils and mayors in the larger cities have gained in stature and resources, however. In the past 10 years, several cities have changed their election system from at large to ward based; separately elected mayors are now more common; and the salaries and resources of the mayors have increased. Politically, local mayors seem to be more powerful actors than they were 20 years ago.

Numerous state-mandated functions—primarily welfare, the court system, jails, and public health—are administered by counties. County governments also perform general governmental functions in unincorporated areas and maintain several extensive regional services such as parks, libraries, and health inspections. Counties generally are influential in coordinating countywide policies that directly affect cities. Important decisions on the location of county facilities, land use, and transportation coordination are county concerns.

In each county, citizens elect five-member boards of supervisors, as well as several independent administrators. Supervisors are powerful officials with considerable influence in all county decisions within their district. They command high salaries and significant staff support. An Orange County supervisor recently referred to himself as "the CEO of my District"; they are prone to focus on the concerns of their district rather than the county as a whole.

Each county also has a transportation agency that is responsible for transportation planning and the operation of transportation carriers. The governing bodies of the agencies consist of the County Supervisors, various appointed city officials, and "public" members selected by the Board of Supervisors.

The region also is governed by several powerful special districts, some of which have significant influence on regional decisions. The South Coast Air Quality Management District (AQMD) develops and implements regional air quality plans, with the approval of state and federal agencies. Its authority includes plans for control of traffic congestion and land use; thus, it is directly involved in basic decisions of local government. The AQMD is governed by a board consisting of 12 representatives, 4 appointed by representatives of the five counties, 4 selected by the cities in the five counties, 1 additional selected by the cities in Los Angeles

County, and 1 each appointed by the governor, the Assembly speaker, and the chair of the Senate Rules Committee.

Regional coordination is assumed by the Southern California Association of Governments (SCAG), a council of governments composed of voluntary members representing cities and counties. The SCAG has a major role in developing the transportation and land use portions of the air plans developed by the AQMD. The SCAG is the designated agency for the Intermodal Surface Transportation Efficiency Act of 1991 (ISTEA). In that capacity, the SCAG has the power to recommend approval of individual projects proposed by cities if they are deemed in conformity with regional plans. The SCAG slowly has expanded its authority, changing from an advisory body to one with limited decision-making authority, particularly in transportation, air quality, and land use planning.

The SCAG is governed by a regional council of 70 elected officials. Sixty-three are selected by elected officials in districts, and seven are supervisors, one from each county and two from Los Angeles. The large governing body represents a recent change, in response to criticism from cities. The relationship between the SCAG and local governments always has been contentious. Criticism of the SCAG by city officials is common, and cities and counties have dropped their membership at various times.

Political traditions and recent history suggest contradictory influences on regional development. Historically, Los Angeles was built on bold regional decisions and large and daring public works projects (Erie, 1992). The regional water system is the consequence of farsighted land purchases, canals, and aqueducts that have diverted water from mountain lakes and the Colorado River. The Metropolitan Water District, a regional special district, continues this tradition and is the major wholesaler of water throughout the region. A large rail network, privately owned but developed with public planning and support, once spanned the area, well before population had reached the outer portions of it. The City of Los Angeles, during the first part of the century, annexed a strip of land leading to the ocean and developed a major port.

Counties have been known for innovative and often daring policies. Los Angeles County pioneered in service contracts during the 1950s. The so-called Lakewood Plan permitted cities to incorporate without significant capital investment and encouraged suburban expansion in eastern and southern portions of the County (Crouch & Dinerman, 1964). Orange and Riverside Counties in recent years won praise for innovations in welfare and social service reform.

Cities also have implemented costly and innovative enterprises. Major redevelopment projects are common in most of the larger and many smaller cities. During the 1970s and 1980s, many cities transformed sleepy old downtowns into regional hubs, and many created major shopping centers and industrial parks. Redevelopment in Los Angeles has been responsible for the large, post-1960 skyline and the building of several facilities for the fine arts. Anaheim, with a population of 260,000, built two major league sports facilities and a sizable convention center. Cerritos, a Los Angeles County city with a population of 65,000, used the redevelopment process to build a major shopping center, several hotels, and a large facility for the performing arts.

Political attitudes remain fundamentally conservative, particularly outside the City of Los Angeles. Large majorities of the delegations in both Congress and the state legislature are Republican, and among those Republicans are some of the nation's most ideologically conservative representatives.

Paul Peterson's *City Limits* (1982) perceptively describes the values that dominate city decision making throughout the region. Peterson argues that concerns for growth and development dominate decision making when cities perceive intense competition for resources with one another and sources of wealth can move readily from one city to the next. Cities in such a setting act like competing firms, using their resources to attract income, paying much less attention to expenditures that supply services to the residents or assist the less fortunate. Major spending efforts aimed at resource acquisition and significant redevelopment activities are common. The presence of numerous cities, a mobile population, diverse business and commercial interests, and a taxing system that permits cities to retain taxable resources encourages these patterns.

The search for new taxable resources expanded when voters approved Proposition 13 in 1978. That initiative limited property tax income in cities, making them more dependent on commercial income. Competition for resources thus increased as each major city used its resources to attract shopping centers, auto malls, and other revenue-producing land uses (Saltzstein, 1986). Since 1978, city officials have focused on internal city concerns, and conflict among cities and between cities and counties has increased. The atmosphere has not been conducive to regional decision making.

Recent civil disturbances in Los Angeles undoubtedly added to the parochial view taken by city officials outside the central city. Los Angeles increasingly is viewed as an alien government, one with problems others hope to avoid. Earthquakes, wildfires, and declines in the economy all act to discourage common obligations to the region as a whole.

The recession, the incorporation of formerly county-controlled areas, and severe reductions in state support reduced county resources significantly in the 1990s. The cumulative effect of these changes and increased service needs of a larger deprived population led to a major fiscal crisis in Los Angeles County in 1995, and proposals were made to reduce the size of government by 20% to cover a deficit of more than $1 billion.

In Orange County, questionable investment practices by the elected county treasurer resulted in a loss of $1.57 billion and the filing of bankruptcy. City, special district, and school funds were included in the loss. City and county leaders, never the best of friends, became quite hostile toward one another as problems associated with the bankruptcy surfaced.

Thus, the characteristics of political organization and values provide little support for the development of regional governments. There is little tradition of central authority in the region. City officials normally look inward, concentrating on resource acquisition. The SCAG has lacked legitimacy in the eyes of many city and county officials. Recent events and economic changes have increased conflicts among governments.

■ III. Regional Decision Making

Regional decisions do not naturally occur in greater Los Angeles. There is little incentive for either county or city governments to develop coordinated policies. Nevertheless, during the late 1980s several regional policies were put into place. Federal and state environmental legislation and an important court case necessitated the creation of a strong regional air quality agency and required issues of air quality improvement to influence land use and transportation decisions. State and local referenda and a new federal law radically increased transportation funds and required new planning mechanisms. Local and state politicians proposed far-reaching legislation for regional governance. SCAG officials responded to a general feeling on the part of many city officials that their actions were unresponsive to the concerns of individual cities by expanding the governing council and broadening participation in its regional plan. When economic and political conditions changed in the 1990s, this policy window closed and attempts were made to limit these initiatives.

A. Air Pollution Control

Legislation often has unintended effects. Certainly no one at the time imagined that passage of the Federal Clean Air Act of 1970 and the establishment of the Environmental Protection Agency would begin the development of regional policies for greater Los Angeles. Federal standards for various kinds of pollutants were mandated by the act with the assumption that cities failing to meet these standards would experience sanctions such as loss of federal funds.

Readings of the air quality in greater Los Angeles revealed that pollution levels significantly exceeded the federal standards for three of the four major sources of pollution. To no one's surprise, the region was labeled as the most polluted in the nation (Howitt, 1984; Cook, 1985).

California legislators approved the Lewis Air Quality Act in 1976 to provide the means to bring the state into compliance with federal standards through the cooperation of local governments. Air Quality Management Districts were created by this act in each urban area that was out of compliance with the federal standards. Each district was expected to develop its own plan to bring the region into conformity with federal and state standards.

The 1977 amendments to the Clean Air Act extended the deadlines for compliance and established a local air planning process, requiring separate plans for each local area that had not met the standards. State plans to bring each affected area into compliance were required (Rosenbaum, 1991, pp. 180-182). In the same year, however, the Coalition for Clean Air, an important environmental interest group, successfully took the state to court for its failure to enforce the federal act. Consequently, federal implementation plans were required in addition.

Over the years, the AQMD has become a powerful organization with an annual budget of more than $100 million and more than 1,000 employees. Much of its revenue comes from fines and inspections. Thus, it is able to act quite independently of other government agencies. Most of the AQMD's authority over "stationary" sources of pollution is direct and clear. It has the power to issue permits and can fine industries that pollute. Its power over "indirect" sources of pollution, however, has been less well defined. Indirect sources are primarily automobiles and trucks. These sources are responsible for about 60% of the air pollution. The transportation system, land use patterns, and the individual habits of drivers are the major variables affecting these pollutants.

State legislation requires the AQMD to develop long-range plans that will lead to conformance with federal and state pollution standards. The plans are updated and revised every 2 years and must be approved by both the state Air Resources Board and the federal Environmental Protection Agency. The 1989 plan was the first to propose a series of measures aimed at indirect sources, and thus local governments became major actors in the abatement process.

The plan is viewed by one expert as the most radical air quality plan in U.S. history (Rosenbaum, 1991, p. 170). It contains 24 compliance measures specifically aimed at local government control of indirect sources of pollution. Although the goal is pollution abatement, the effect is a regional plan that more rationally coordinates land use, transportation systems, and personal habits to reduce automobile usage and other sources of pollution.

Many of these proposed measures assume that local governments will expand their authority over the private sector and implement measures that would alter the traditionally cozy relationship between public management and the business community. Land use decisions must be related to reductions in automobile traffic. Government's power to regulate must be used to encourage innovations in business practice that discourage traffic during peak hours. Some household activities, such as the burning of barbecue fluid and the use of gasoline lawn mowers and swimming pool heaters, will be limited. Local officials therefore are expected to enforce regional policies that may conflict with local self-interest.

The extent to which local governments have to comply with these measures is less than clear. The SCAG maintains that "If local governments do not make reasonable progress in implementing control measures for which they are responsible, the AQMD has the legal authority to assume some of these responsibilities" (Southern California Association of Governments, 1993). Ultimately, sanctions may be imposed on the area if meters registering the amounts of pollutants in the air indicate that federal or state standards are not met. Many factors contribute to air pollution. Compliance with restrictions, for example those mandating carpooling, therefore may not clearly contribute to reducing the amount of pollutants in the air.

Compliance activities occurred in many cities. A study of local government response to the efforts of the AQMD in 1992 revealed that one-third of the cities had implemented many of the control measures (Saltzstein, 1992, 1994). Thus, the AQMD has become a significant influence on important local government decisions.

Political changes on the AQMD board and a heightened concern about the effect of environmental controls on job retention has led to a harsh reaction to the power of the AQMD since 1992. City officials became increasingly concerned about their representation on the AQMD board. Orange County officials went as far as attempting to remove the board's chair, Henry Wedaa of Yorba Linda, in 1994. Wedaa had become an extremely influential spokesperson for Orange County interests but was viewed by many as too environmentally conscious. The retirement of Mayor Tom Bradley in 1994 removed one of the most vigorous and visible spokespersons of environmental concerns. State appointees in recent years also have been decidedly more conservative than in the past. The changes in the political climate have had an effect on the AQMD's attempt to influence city policies. Since 1991, AQMD officials have turned their attention toward market incentives rather than command and control strategies, and the enforcement of local government compliance measures has been less vigorous. Thus, the importance of the AQMD as a regional influence on local governments is an open question at this point in time. Certainly the expansion of power, much evident in 1989, has been curtailed.

B. Transportation Planning

Los Angeles and its freeway system are closely identified. Congestion is assumed, and alternatives to driving alone usually are thought to be impossible to implement given the frequency of long commutes and the alleged love affair between Angelenos and their cars. In the 1970s, however, most freeway construction was halted. Transportation planners and interests at the state level assumed that more freeways would create an inefficient transportation system and hurt the environment.

In the late 1980s, a new approach was proposed, linking transportation policies with environmental concerns. Freeway construction would be dependent on approved plans for congestion management. These plans would take into consideration the link between land use and auto traffic. Thus, transportation plans were required to coordinate with land use and environmental plans. Provisions had to be made to limit future use of freeways by coordinating jobs, housing, and commerce. Transportation agencies then became a more important part of the regional planning process: Transportation funding had to be related to land use, so regional transportation agencies became important players in land use decisions.

The expanded role of transportation agencies in the regional planning process resulted from a combination of county referenda and state laws and referenda permitting the consolidation of county agencies. Despite the obvious complexity of these changes, most were accomplished within a 3-year period. Proposition 111, approved by voters in 1990, dedicated a 9 cent increase in the gas tax for local transportation contingent on congestion management plans. Initiatives in individual counties in the late 1980s also increased local sales taxes by 0.5% for transportation purposes and required stronger congestion controls. New federal legislation, the Intermodal Surface Transportation Efficiency Act (ISTEA), designated an estimated $1.36 billion for transportation to the region and also required links between transportation, air quality, and the environment (Bollens, 1993). A Metropolitan Planning Organization (MPO), in this case the SCAG, was given considerable discretion in how these funds were distributed. The influx of funds encouraged an ambitious program of transportation improvements and widened the horizon of possibilities proposed and debated by transportation agencies. Los Angeles County embarked on an extension of its public transportation system, including a subway between the downtown area and Hollywood and a trolley between the civic center and Long Beach. Orange and Riverside Counties built tollways to relieve the most congested corridors. Bold new initiatives were proposed, including high-speed trains, monorails, and automated roadways.

Significant institutional changes also followed. In both Los Angeles and Orange Counties, formerly autonomous countywide agencies for transportation planning and the bus system were combined, creating the Metropolitan Transit Authority (MTA) in Los Angeles and the Orange County Transit Authority (OCTA). The SCAG's role in transportation expanded considerably because it was designated as the agency responsible for determining the allocation of ISTEA funds. Because the federal legislation was designed to give local agencies more flexibility, this authority was a considerable extensions of the SCAG's power.

Both of these changes added to the complexity of the regional process. The new transportation agencies, with enhanced sources of funds and influence over land use decisions, became more important actors. The SCAG's expanded role was met with hostility from Orange County, and legislation was proposed to designate Orange County as a separate MPO.

To receive funds for local street and road improvements from the transportation authorities, city officials must develop plans that will lead to limiting automobile usage, a constraint very similar to the AQMD requirements. The difference is that cities receive money from the transportation districts and negative sanctions from the AQMD. Hence, the transportation process is viewed more favorably by city officials than is air pollution control. Orange County cities have been trying to develop common plans for both agencies through a subregional organization that also has served as a representative of the common interests of cities and the county in regional forums.

The mid-1990s witnessed some retrenchment in metropolitan transportation planning. Bankruptcy in Orange County resulted in the layoff of key staff members who were coordinating the city and county plans. Thus, the subregional body has been dormant since December of 1994. Some have proposed the use of some of the transportation funds to pay off part of the debt associated with the bankruptcy. Cost overruns and difficulties with the tunneling of a new subway in the Hollywood area have raised questions about the MTA's role as a regional actor.

C. Regional Governance

From 1985 to 1990, political leadership in California devoted much time and effort to the creation of regional governance for greater Los Angeles. Large sums were raised to study the governmental problems of the region, hearings were held across the state, and several complex bills were introduced in the state legislature. The mayor of Los Angeles took the lead with a major study commission. The governor announced that growth management would be a major part of his agenda. The speaker of the Assembly, next to the governor the most powerful state official, was the author of one of the proposals, and the chair of the Local Government Committee in the State Senate, a respected and powerful Orange County politician, put together a major study of the region's political problems and held hearings throughout the state.

The Bradley Administration in Los Angeles had added much wealth to the city through major redevelopment efforts in the 1970s and 1980s. The increased traffic, pollution, and inconvenience was creating much unease among the City Council members and within parts of the community. To deal with the array of problems presented by changes in the city generally and also to answer some of the criticism, Bradley appointed the Los Angeles 2000 Committee and secured more than $1.25 million from private sources for a planning process and a strategic plan. Significant elements of the business and academic communities as well as community activists and politicians were involved in the process (Trombley, 1988).

The final report, released in 1988, called for two powerful regional agencies: a Regional Growth Management Agency to set policy for land use, housing, and transportation, and a regional environmental agency merging the AQMD with agencies that manage water control and solid waste. It also advocated an airport authority coordinating the operations of all existing airports. Massive increases in public investment also were proposed.

Extensive activity at the state level occurred in 1991. The LA 2000 report and a similar effort in the San Francisco Bay area provided the backdrop for a flurry of activity from Governor Pete Wilson and members of both houses of the legislature. The governor formed a Growth Management Council consisting of several of his cabinet members and held hearings throughout the state (Trombley, 1991). Senate and Assembly committees also held hearings. Five major legislative bills were drawn up in both houses and by members of both parties. They ranged in comprehensiveness from the creation of a unified, strong metropolitan government to a voluntary agency to fund infrastructure improvements. Within these bills was much room for compromise. The debate suggested bipartisan support for significant reform.

This "window of opportunity" for metropolitan governance closed rapidly. As the economy soured and the governor's standing in the polls dropped, his interest in growth management waned. Legislative action turned to the more pressing needs of the state budget in a time of declining resources. Mayor Tom Bradley's interests were deflected by more immediate political problems and a run for the governorship. His replacement, Richard Riordan, has yet to articulate a position on growth management and metropolitan governance.

Generally, bad times changed the political debate. In 1991, growth management was viewed as a way of encouraging economic growth. Traffic congestion, foul air, and water shortages were seen as hurting California's competitive position. Stronger regional governments were seen as a means of improving the state's competitive position. As the economy soured, political leaders argued that the public sector was hurting the economy through coercive taxation and regulation. Support for regional solutions among politicians and private-sector leaders of interest groups seemed to be nonexistent.

Currently, the SCAG is in the midst of developing a general plan for the region. Participation in development of the plan has been broadened through the SCAG's new governing process, extensive public hearings, and a concerted attempt to secure diverse forms of participation. The plan itself, however, is best described as a guide for possible courses of action because the SCAG remains fundamentally an advisory body (Fulton & Newman, 1992).

■ IV. Toward Regional Government?

Social, economic, and political trends in greater Los Angeles indicate that the preconditions for serious metropolitan decision making are not present. The area has become more dispersed and more diverse in recent years. As wealth has declined and general funding sources have become more limited, cities and counties have turned inward in their worldview to search for new resources and protect themselves from their neighbors. Leadership has avoided metropolitan issues in fear of a public backlash. Regional districts are a frequent target of political campaigns. Regulation often has been viewed as a cause of the loss of employment opportunities.

Some form of metropolitan coordination nevertheless has taken place. Cities and counties have responded to proposals from the AQMD, the SCAG, and the transportation authorities. Measures encouraging congestion management, auto trip reduction, and new, less polluting technologies have been adopted by many cities. Regional plans are a part of the public agenda.

Opportunities for regional decisions in greater Los Angeles occur infrequently and may end rather abruptly. In Kingdon's words, "[P]olicy windows open infrequently and do not stay open long" (1995, p. 166). Windows open when political conditions, resources, and leadership converge. All these conditions were present in 1991. By 1993, the window had closed. The late 1980s and early 1990s expanded the power of regional agencies and forced some coordination among cities and counties. Less successful were attempts to get the regional agencies themselves to agree on approaches to public problems. This should be the goal of the next confluence of leadership, resources, and public concern.

As a consequence of the actions taken in the late 1980s, greater Los Angeles has cleaner air, impressive transportation policies, and programs aimed at coordinating land uses for public purposes. Regional decisions, however, are made by single-purpose agencies that are governed indirectly by the electorate and make decisions out of the public limelight.

Is this a good way to proceed? Noted city planners Fulton and Newman find virtue in the status quo.

The truth is that regional planning in Southern California has become a floating crap game. It's not controlled by SCAG or any other single agency. Rather it is centered—if that is the word—in scattered meetings and negotiations and skirmishes that occur over this 100 square mile region. It's not quite what the advocates of metropolitan planning had in mind back in the sixties but in the long run this free-floating system may prove workable for such a fragmented region. (Fulton & Newman, 1992)

There are problems, however, with proceeding in this manner. Regional policies are concerned with tradeoffs between fundamental values. Can we pursue cleaner air and job growth? Do we prefer less freeway congestion or more personal freedom? A jobs/housing balance may require less preservation of open space. The present system often hides these conflicts. Single-purpose agencies give secondary concern to the goals of other agencies. An indirectly elected governing body lacking a strong media presence will not inform the public of its deliberations. Thus, these important conflicts will not be discussed openly. A strong regional government with influence over the special districts and a clear link to the electorate would be better able to discuss the choices that must be made and generate support for them. Perhaps these concerns will become the forum of the next opening of the policy window.

REFERENCES

Bollens, S. A. (1993). *Metropolitan transportation governance in Orange County.* Unpublished manuscript, Department of Urban and Regional Planning, University of California, Irvine.

Cook, K. F. (1985). Pollution control and intergovernmental relations: Is the federal stick necessary? In D. Judd (Ed.), *Public policy across states and communities* (pp. 91-108). Greenwich, CT: JAI.

Crouch, W., & Dinerman, B. (1964). *Southern California Metropolis.* Berkeley: University of California Press.

Erie, S. P. (1992). How the West was won: The local state and economic growth in Los Angeles, 1880-1932. *Urban Affairs Quarterly, 427*(June), 519-554.

Fulton, W., & Newman, M. (1992). When COGS collide: Councils of governments, Southern California Association of Governments. *Planning, 58*(1), 9-15.

Howitt, A. M. (1984). *Managing federalism: Studies in intergovernmental relations.* Washington, DC: CQ Press.

Kingdon, J. W. (1995). *Agendas, alternatives and public policies.* New York: HarperCollins.

Lockwood, C., & Leinberger, C. B. (1988, January). Los Angeles comes of age. *Atlantic Monthly,* pp. 43-52.

Peterson, P. E. (1982). *City limits.* Chicago: University of Chicago Press.

Rosenbaum, W. (1991). *Environmental politics and policy* (2nd ed.). Washington, DC: CQ Press.

Saltzstein, A. (1986). Did Proposition 13 change spending priorities? In T. N. Clark (Ed.), *Research in urban policy 2A* (pp. 145-158). Greenwich, CT: JAI.

Saltzstein, A. (1992, March). Is clean air really worth all of this?: Compliance with the Clean Air Act in greater Los Angeles. Western Political Science Association, Pasadena, CA.

Saltzstein, A. (1994, March). Los Angeles mayors and regional politics: The strains of imposed regional government. Western Political Science Association, Portland, OR.

Silverstein, S. (1988, May 14). Where the jobs are. *Los Angeles Times,* Part C, p. 24.

Soja, E., Morales, R., & Wolff, G. (1984). Urban restructuring: An analysis of social and spatial change in Los Angeles. *Economic Geography, 60,* 195-230.

Southern California Association of Governments. (1993). *Draft regional plan.* Unpublished manuscript.

The next Los Angeles: Turning ideas into action. (1994, July 17). *Los Angeles Times,* Part 5, p. 2.

Trombley, W. T. (1988, November 16). Two agencies proposed to set regional policy. *Los Angeles Times,* Part 1, A1.

Trombley, W. T. (1991, September 15). Wilson tries to map the future of a growing state. *Los Angeles Times,* Part B, p. 1.

4

St. Louis: A Politically Fragmented Area

DONALD PHARES
CLAUDE LOUISHOMME

■ Whither Regional Governance?

The political and fiscal organization of local government in the context of effective regional governance is a topic of considerable scrutiny and concern. This chapter examines the St. Louis region, which has a long tradition as a textbook example of a fragmented urban governance and fiscal structure and the associated litany of problems. As one major indicator, fewer people are served per local jurisdiction in St. Louis than in any other major urban area except Pittsburgh. The range is from more than 44,000 in Baltimore to 3,554 in St. Louis and 3,063 in Pittsburgh (Confluence St. Louis, 1987, p. 18). The entire 12-county, 2-state (Missouri and Illinois) metropolitan area (MSA) had 771 local jurisdictions as of the 1992 *Census of Governments.* This is an increase of 140 governments (22%) since 1972; 101 are in Missouri. Virtually all (133) are special districts, and 71% of those (94) are in Missouri. As will be discussed later in the chapter, attempts to deal with regional issues in the St. Louis region often have been manifested as new, single-purpose, special districts.

Although taking a literal stance against fragmentation per se would be without basis, there are substantive problems associated with the fragmentation, uncoordinated regional governance, and self-interest-driven ad hoc annexations and incorporations that have characterized the St. Louis area. The development and well-being of this region is far too important to predicate governance and fiscal arrangements on an ideological stance. In addition, the issues are far too complex to assume that the laissez-faire approach of depending solely on a political free market will lead to governance appropriate to the St. Louis urban region. As the following text will discuss, the tension between "local" and "regional" continues to play out.

■ The St. Louis Area "Functional City"

To paraphrase a definition of the "functional city" provided by Savitch and Vogel earlier in this volume, a region is a functional city when traditional American cities extend themselves and interact with areas around the urban core. We define the "Functional City" as the combined area of the City of St. Louis (which, it should be noted, is a county as well) and St. Louis County (see Figure 4.1). This is highly defensible for the reasons given below. For ease of subsequent discussion, we refer to the City of St. Louis as the "City," St. Louis County as the "County," the combined City and County as the "Functional City," and the 12 counties (7 in Missouri, 5 in Illinois) in the census definition as the MSA.

The Functional City accounts for the overwhelming proportion of residents and jobs in the St. Louis region. Although it represents only 2 of the 12 counties in the MSA, it represented nearly 85% of the Missouri portion of the region's population in 1970 and 75% in 1990. The Functional City also contains the overwhelming share of the region's employment and economic base.

In addition to serving as the center of population and economic activity for the region, the Functional City also has a long history of being the focal point for what regional cooperation and governance does exist. Although quite modest, it is unparalleled in the context of the larger urban region.

A third and powerful factor leading to this definition of the Functional City is geographical in form but truly functional in impact. The Mississippi and Missouri Rivers separate this area from most of the other counties in the MSA (see Figure 4.1). The Mississippi River serves both as a symbolic and as a functional and legal dividing line between Missouri and Illinois. The 5 Illinois counties that are part of the MSA are in close proximity to the Functional City, but the state boundary, the Mississippi River, represents a real demarcation that exacerbates problems of cooperation. Differences in history and tradition, laws and practices, institu-

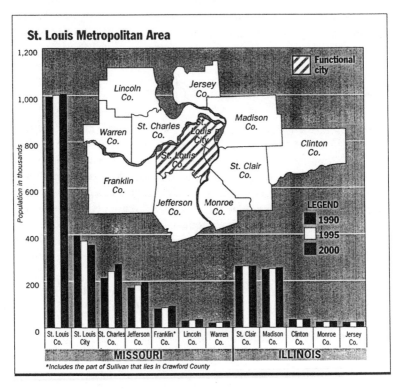

Figure 4.1.   Map 1: St. Louis Metropolitan Area
SOURCE: American Statistical Association, St. Louis chapter.

tional arrangements, and interests for the two states inhibit cooperation, and difficulties are reinforced by geography.

Whereas the Mississippi River represents a barrier to *inter*state cooperation, the Missouri River serves as a similar demarcation for *intra*regional endeavors, especially between St. Louis and St. Charles Counties. In this instance, however, it is not simply a boundary that inhibits cooperation. The history and development trajectory that the Functional City area has manifested for more than 100 years contrast markedly to the history and development pattern in contiguous St. Charles, Jefferson, and Franklin Counties.

Until the 1970s, these counties, which surround the Functional City, primarily were rural farming communities. The building boom of the 1980s, however, was particularly concentrated in St. Charles County. That boom involved commercial, industrial, and residential construction. Growth in the outlying counties has continued. Jefferson County, which borders to the south, also has seen a tremendous growth in its population in the last two decades, despite the fact that it is characterized by a terrain much less hospitable to development.

Although St. Charles, Franklin, and Jefferson Counties have been dealing with rapid growth (but on a small base) in the last two decades, St. Louis County has reached a plateau in its population and its share of the region's economy. Within this context, it has begun to deal with some of the same issues that have vexed St. Louis City for decades, such as an aging housing stock, obsolete commercial and industrial facilities, and a pool of structurally unemployed workers.

It should be stressed, however, that although both St. Louis City and St. Louis County are dealing with similar types of conditions, the scale of the problems is still much smaller in the County. The City has experienced a massive outflow of residents, businesses, and jobs since its peak in the 1950s. The County has not yet experienced such losses.

■ A Profile of the St. Louis Area

The St. Louis MSA's population (Missouri portion) has remained stable over the past two decades (see Table 4.1). This overall stability, how- ever, has been coupled with significant shifts from the region's historical core (the City of St. Louis) to outlying counties and communities such as the City of St. Charles, Franklin County, and Jefferson County. Between 1970 and 1990, the population was virtually static at 1.85 million, yet during the same period, the number of people living in the Functional City dropped from 1.57 to 1.39 million (see Table 4.2). Most of this loss occurred in St. Louis City.

In addition to having a static population, and despite the dramatic loss experienced by the City, a growing percentage of the area's residents live in an urban setting, some 91% in 1990 compared to 87% in 1970. The region's rural population declined from more than 12% in 1970 to less than 10% in 1990. The Functional City itself is virtually entirely urban (Table 4.2).

TABLE 4.1 Basic Socioeconomic Characteristics of the St. Louis Metropolitan
             Statistical Area,[a] 1970 and 1990

	1970		1990	
	Number	Percentage	Number	Percentage
Population				
Total	1,854,647	100.0	1,855,104	100.0
Urban	1,623,234	87.5	1,680,804	90.6
Rural	231,413	12.5	174,300	9.4
Gender				
Male	882,551	47.6	885,149	47.7
Female	972,096	52.4	969,955	52.3
Race				
White	1,544,603	83.3	1,493,063	80.5
Nonwhite	310,044	16.6	362,041	19.5
Age (years)				
0-21	761,655	41.0	577,055	31.2
22-44	521,762	28.0	691,971	37.4
45-64	383,313	21.0	351,184	18.9
65 and over	187,917	10.0	234,894	12.5
Education (persons 25 years of age and over)				
Less than 12th grade	519,533	51.3	279,463	23.3
High school graduate	290,746	28.7	344,100	28.5
Some college	92,538	9.1	303,379	25.3
4 years of college or more	110,306	10.9	274,012	22.9
Top employment sectors				
Manufacturing	205,763	24.3	111,245	12.3
Wholesale-retail trade	156,225	18.5	199,581	22.1
Durable goods	127,413	15.1	62,740	7.0
Professional and related	72,598	8.6	144,007	16.0
Educational services	50,845	6.0	84,441	9.4
Finance, insurance, etc.	58,232	6.9	66,623	7.4
Persons below poverty level				
Total	192,591	100.0	178,937	100.0
65 and over	41,816	21.7	22,533	12.6
Under 65	150,775	78.3	156,404	87.4

SOURCE: U.S. Census of Population; 5th count data tapes for 1970, STF3 data tapes for 1990.
a. Missouri portion of the metropolitan area (MSA) only.

This outward population migration has resulted in the spread of conditions usually associated with the urban core. For example, the St.

TABLE 4.2 Basic Socioeconomic Characteristics of the St. Louis Functional City Area,[a] 1970 and 1990

	1970 Number	1970 Percentage	1990 Number	1990 Percentage
Population				
Total	1,573,589	100.0	1,390,214	100.0
Urban	1,534,067	97.5	1,372,580	98.7
Rural	39,522	2.5	17,634	1.3
Gender				
Male	743,418	47.2	654,454	47.1
Female	830,171	52.8	735,760	52.9
Race				
White	1,267,621	80.6	1,038,879	74.7
Nonwhite	305,968	19.4	351,335	25.3
Age (years)				
0-21	636,626	41.0	414,866	29.8
22-44	436,492	26.0	510,062	36.6
45-64	335,460	22.0	269,126	19.4
65 and over	165,011	11.0	196,160	14.2
Education (persons 25 years of age and over)				
Less than 12th grade	440,642	50.7	211,916	23.0
High school graduate	243,402	28.0	243,851	26.6
Some college	82,893	9.5	228,442	25.0
4 years of college or more	102,446	11.8	132,485	25.4
Top employment sectors				
Manufacturing	170,452	23.7	48,105	7.2
Wholesale-retail trade	136,812	19.0	147,282	22.0
Durable goods	102,573	14.3	71,542	10.7
Professional and related	66,563	9.3	115,752	17.3
Educational services	44,418	6.2	55,388	8.3
Finance, insurance, etc.	52,446	7.3	86,753	13.0
Persons below poverty level				
Total	169,184	100.0	149,874	100.0
65 and over	34,957	20.7	19,333	12.9
Under 65	134,227	79.3	130,541	87.1

SOURCE: U.S. Census of Population; 5th count data tapes for 1970, STF3 data tapes for 1990.

a. The functional city comprises St. Louis City and St. Louis County.

Charles area experienced explosive growth in residential and commercial development. A substantial amount of the new commercial development is located on either side of a primary east-west interstate highway, I-70. The shopping malls and commercial strips that now line the I-70 corridor occupy land that was used for farming as recently as 10 years ago. With these changes have come high-density subdivisions, overcrowded streets and highways, inadequate public infrastructure, and strong demands on public services such as schools.

In addition to the intraregional shifts in population that have occurred in the last two decades, the racial composition of the region also has changed, albeit very slightly. The region remains characterized by two primary racial groups: whites and blacks. In 1970, 83% of the MSA residents were white; by 1990, the proportion had dropped slightly, to 81%. Between 1970 and 1990, the area population went from 16% to 18% black. Other minorities in the region have more than quadrupled between 1970 and 1990, but they still only number about 28,000. The Functional City has seen a larger increase in the proportion of nonwhite residents (19% to 25%), but this has occurred primarily through out-migration.

Between 1970 and 1990, the region's population grew older. Whereas in 1970, 41% was under 21 years of age, about one-half between 22 and 64, and only 10% over the age of 65, by 1990 persons 21 and younger represented only 31.2%, age 22 to 64 had grown to 56.3%, and those over 65 were almost 13%. The Functional City mirrors this regional pattern.

In these two decades, the region became better educated. Whereas in 1970 more than 51% of the population had less than a 12th grade education, by 1990 that proportion was only 23%. At the other end of the educational spectrum, there was a significant increase in the percentage of the population that had completed 4 years of college or more: It more than doubled, from 11% in 1970 to 23% by 1990. The Functional City showed an almost identical trend.

St. Louis area employment traditionally has been heavily dependent on manufacturing and transportation; however, the national shift of America's economy to a service orientation is clearly manifest. For example, in 1970 manufacturing (24%) was the leading sector, followed by wholesale/retail trade (19%) and durable goods (15%). By 1990, wholesale/retail trade represented the largest industry category (22%), and manufacturing had fallen to 12.3%. The growing importance of the service industry is reflected in the fact that in 1970, services constituted only 22% of employment, but by 1990 this proportion had increased to 33%. This trend is pronounced in the Functional City as well. Manufacturing fell from 24% to 7%, services rose from 23% to 38%.

Over the last 20 years, income for both the St. Louis area and the Functional City has risen. Despite this increase in income, however, the number of persons below the poverty level remained fairly constant, falling 7%, from 192,591 in 1970 to 178,937 in 1990. It should be noted that 84% of the region's poor were in the Functional City in 1990 but also that its percentage drop was far greater than for the MSA (11% versus 7%).

■ St. Louis as a Case Study

St. Louis voters approved the legal separation of the City from its county in 1876, when there were only 5 cities in the entire county. By 1936, there were 28, and by 1955, the number had jumped to 96 and the Functional City dominated the urban region. These new cities often began as subdivisions that were incorporated for control over planning, zoning, land use, and the tax base. Later on, intergovernmental funding became an added incentive for incorporation.

Commenting on the situation that had evolved by 1957, the Metropolitan St. Louis Survey noted the following "continuing problems" for the Functional City area (1957, pp. 83-85). Given the passage of almost four decades since its report, the summary reflects an amazingly prescient account of the present St. Louis region.

- Wide public service disparities,
- Substantial variation in ability to finance governmental services,
- Inadequate essential areawide services,
- Failure to recognize responsibility to the urban region,
- Competition among cities for tax base,
- Growth occurring in unincorporated St. Louis County with ad hoc annexations and incorporations not providing an adequate solution, and
- The County needing to provide an increasing amount of municipal services to residents in unincorporated areas.

By the 1960s, municipal development in the County reached the haphazard pattern depicted in Figure 4.2. This trend would have continued but was stopped in 1963 when the Missouri Supreme Court (*City of Olivette v. Graeler*) ruled that the impact of new annexations and incorporations on *all of the County* must be taken into account. The vested

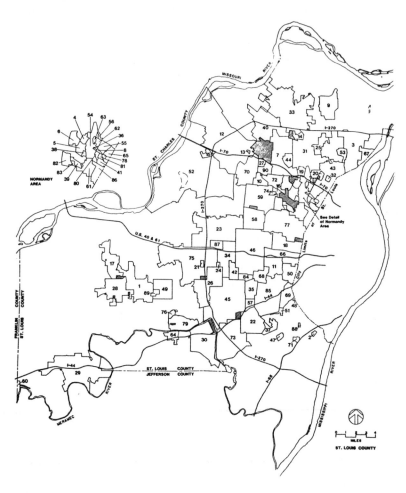

Figure 4.2. Map 2: Municipalities

1. BALLWIN
2. BELLA VILLA
3. BELLEFONTAINE
   NEIGHBORS
4. BELLERIVE
5. BEL NOR
6. BEL RIDGE
7. BERKELEY
8. BEVERLY HILLS
9. BLACK JACK

10. BRECKENRIDGE
    HILLS
11. BRENTWOOD
12. BRIDGETON
13. BRIDGETON TERRACE
14. CALVERTON PARK
15. CHAMP
16. CHARLACK
17. CLARKSON VALLEY
18. CLAYTON

19. COOL VALLEY
20. COUNTRY CLUB HILLS
21. COUNTRY LIFE ACRES
22. CRESTWOOD
23. CREVE COEUR
24. CRYSTAL LAKE PARK
25. DELLWOOD
26. DES PERES
27. EDMUNDSON
28. ELLISVILLE

29. EUREKA
30. FENTON
31. FERGUSON
32. FLORDELL HILLS
33. FLORISSANT
34. FRONTENAC
35. GLENDALE
36. GLEN ECHO PARK
37. GRANTWOOD
    VILLAGE

▨ Unincorporated areas surrounded by municipalities

interest of one (small) group would now be viewed in the context of the impact on the broader region.

This status prevailed for two decades. Although it did not stop annexations or incorporations or promote regionalism, it did curtail the earlier trend. In 1983, however, the Missouri Supreme Court reversed its 1963 decision (*City of Town and Country v. St. Louis County, et al.*), thus allowing once again far greater freedom for annexations and incorporations. This new legal environment elicited "land rush" activity to capture tax base in the rapidly developing unincorporated portions of the County. Barely lip service was given to emerging regional issues.

■ Regional Cooperation or More Fragmentation: The Power of the Fisc?

Ninety cities now exist in St. Louis County (Figure 4.2; the shaded areas are unincorporated pockets). They provide municipal services to more than 600,000 County residents (over 60%) in 254 square miles. The unincorporated county area contains almost 400,000 people in 268 square miles. They receive municipal services from St. Louis County government. Several issues result from this political and fiscal organization.

First, the County has one of the largest responsibilities for municipal services in the state of Missouri. Second, population in these municipalities ranges from 11 to more than 50,000; 2 have less than 100 people, 21 fewer than 1,000, 20 more than 10,000, and only 9 more than 20,000.

Third, the resource base to support services is distributed very unevenly. This results in vast differences in both the quantity and the quality of services and often forces the necessity of compromise. Resource disparities are further exacerbated by the annexation and incorporation

38. GREENDALE	52. MARYLAND HEIGHTS	66. RICHMOND HEIGHTS	80. VELDA VILLAGE
39. HANLEY HILLS	53. MOLINE ACRES	67. RIVERVIEW	81. VELDA VILLAGE
40. HAZELWOOD	54. NORMANDY	68. ROCK HILL	HILLS
41. HILLSDALE	55. NORTHWOODS	69. SHREWSBURY	82. VINITA PARK
42. HUNTLEIGH	56. NORWOOD COURT	70. ST. ANN	83. VINITA TERRACE
43. JENNINGS	57. OAKLAND	71. ST. GEORGE	84. WARSON WOODS
44. KINLOCH	58. OLIVETTE	72. ST. JOHN	85. WEBSTER GROVES
45. KIRKWOOD	59. OVERLAND	73. SUNSET HILLS	86. WELLSTON
46. LADUE	60. PACIFIC	74. SYCAMORE HILLS	87. WESTWOOD
47. LAKESHIRE	61. PAGEDALE	75. TOWN & COUNTRY	88. WILBUR PARK
48. MACKENZIE	62. PASADENA HILLS	76. TWIN OAKS	89. WINCHESTER
49. MANCHESTER	63. PASADENA PARK	77. UNIVERSITY CITY	90. WOODSON TERRACE
50. MAPLEWOOD	64. PERRLESS PARK	78. UPLANDS PARK	
51. MARLBOROUGH	65. PINE LAWN	79. VALLEY PARK	

activity mentioned earlier. The richest unincorporated areas are being absorbed, often leaving poorer areas (dominantly residential) to turn to County government, fend for themselves, or go without services.

As one manifestation, per capita taxable property (1993) ranged from $2,178 to $143,285 (the County average is $12,366). This reflects how property wealth is concentrated in certain areas. Although taxable property variation contributes in part to fiscal problems, it helps more to underscore the problem, because it accounts for only 8% of municipal revenue and 11% of taxes. A much more crucial factor for municipal finance relates to the local sales and gross receipts (on utilities) taxes and how they are distributed among cities. These revenues now account for almost 57% of all revenue and more than 75% of taxes for the County's municipalities.

In 1977, voters approved a one cent countywide sales tax. The proceeds were distributed to cities in two ways, "point of sale" and "pool." Point of sale allowed cities that already had a sales tax to keep all revenue generated within their boundaries. Pooling provided a per capita distribution of all revenue collected in cities without a sales tax prior to 1977, plus all funds collected from unincorporated St. Louis County.

After the 1983 *Town and Country* decision mentioned earlier, new annexations burgeoned. Cities looked to annex contiguous areas with a commercial/industrial base that would augment existing revenues. The issue of sales tax distribution took on new importance under this decision, because loss of wealth in unincorporated areas as a result of annexation led directly to a reduced tax base for all pool cities as well as for County government. To help mitigate this, the Missouri legislature passed legislation in 1983 placing all newly annexed and incorporated areas in the pool allocation. They would receive a per capita allocation and not be able to retain all sales tax receipts.

Sales tax allocation continues to be a major fiscal concern. In 1993, the earlier 1977 allocation formula was revised to adjust for inequities that had become evident. The new allocation provided for a "progressive, sliding scale" sharing of point-of-sale revenues with pool cities. This compensated for some of the wide disparity in sales tax revenue that had evolved. As the Boundary Commission of St. Louis County (1993, p. 11) noted, "Wide disparities in municipal revenue, service needs and tax burdens and the insecure financial future of some cities make it difficult for them to grow."

As of 1993, sales tax receipts continue to differ widely, from a pool allocation of about $75 per capita to a point-of-sale range exceeding 10 times this amount. This has been a very powerful fiscal and political incentive for point-of-sale cities to maintain the status quo; change would mean a loss of revenue.

Another major contributor to municipal coffers, and also to prevailing fiscal tension, is the gross receipts tax on utilities (water, gas, electric, and telephone). A strong incentive exists to capture this revenue base within an incorporated area because all proceeds are kept by a city. Generally, areas with high sales tax yields also tend to have large utility receipts. Per capita utility consumption (1993), the gross receipts tax base varied from a low of about $500 to more than 10 times this amount.

A recent wrinkle on these fiscal machinations arose during the 1994 legislative session. Missouri voters had approved riverboat gaming in a 1993 vote. As a part of this legislation, incorporated areas in which gaming facilities located would receive 2% of gaming proceeds plus $1 per admission. This was seen as a potential major new source of funds, and new legislation was introduced to place all gaming revenues into a countywide pool, allocated on a per capita basis, rather than let it accrue to the city in which the facility was located. Although this revenue pooling scheme was not implemented, it is not yet legislatively "dead." Resource disparities resulting from fragmentation and the lure of new funds keep it alive.

As the above discussion attempts to document, any concerns over local government structure and regional issues in St. Louis have been overpowered by myopic, but perhaps understandable, local fiscal incentives to capture taxable resources. Fourteen annexation proposals were on the November, 1994, ballot; more are in progress. Four new incorporations are being developed, with two on the ballot early in 1995. The land rush goes on.

■ Regional Versus Local Governance: An Ongoing Tension

Two Advisory Commission on Intergovernmental Relations reports (ACIR, 1988, 1992) once again brought into national focus the question of government organization in metropolitan areas. These ACIR reports documented in great detail the informal arrangements in two highly fragmented urban regions, St. Louis and Allegheny Counties. Working in a "public choice" framework, the reports praised the creativity of "public entrepreneurs" and part-time officials in small cities. Reading these reports leaves one with the impressions that this is how regional govern-ance should operate, smaller and more is better, and a public market operating without concern for regional implications works well.

Release of the St. Louis ACIR (1988) report in September coincided with the formal submission of a plan for comprehensive governmental and fiscal reorganization within the Functional City area. There is a stark contrast between the ACIR findings and the provisions set forth in the plan submitted for areawide reform, *Plan for Governmental Reorganiza-tion in St. Louis & St. Louis County* (Board of Freeholders, St. Louis City/County, 1988a).

The ACIR report praised existing fragmented governance arrange-ments, noting in conclusion: "The experience of the St. Louis area in metropolitan organization *has much to teach the rest of metropolitan America*" (emphasis added; ACIR, 1988, p. 168). In sharp contrast, the Board of Freeholders (1988a) proposed major reorganization and fiscal restructuring for all County, municipal, and fire and emergency medical services (EMS) plus the establishment of two new regional governance entities. It is interesting that despite ongoing knowledge of each other's work and staff interaction, the ACIR and the Freeholders came to hold almost diametrically opposite views of what needed to be done. The ACIR view fell into a pro-status quo, public choice framework, whereas the St. Louis Board's plan clearly articulated the need for fundamental reform.

This board and the underlying premises that guided formation of its complex plan will be discussed in the following sections. Focus will be on government organization within the County, because this lies at the core of the debate over provision of public services in the region. Attention will also focus on events since this board went out of existence. The intervening years have been far from uneventful, and the issues of gov-ernance and fiscal structure have become even more contentious.

■ Governmental Reform Using the Missouri Constitution

Missouri's constitution (Article VI, Section 30) is somewhat unique in allowing for the establishment of a board with authority to address issues pertaining to local governance within the Functional City, subject to voter approval. In 1987, the first board in 28 years was sworn in, and it chose the constitutional option "to formulate and adopt any other plan for the partial or complete government of all or any part of the city and the county." Under the purview of this option, the board delved into the myriad issues pertaining to local governance and finance within the Functional City. Although the constitution did allow proposals for restructuring any or all local government, including school and special districts, the board made a conscious decision to focus only on County, municipal, fire/EMS, and regional issues.

■ Toward Comprehensive Governance Reform in St. Louis

The board's focus on municipal/County government organization and finance derived directly from the fragmented governance and fiscal environment that had evolved over decades, as discussed earlier. Context for its work was provided by several local studies. One posed an intriguing question for the title of its final report: *Too Many Governments?* (Confluence St. Louis, 1987).

After lengthy study, the board defined several premises to guide formulation of its plan. The first was complete incorporation to equalize service provision for nearly 400,000 residents in unincorporated areas. Second, County government would provide no municipal services; this responsibility would rest with newly created cities. Third, a governance structure was proposed in which all residents of the County would have access to adequate municipal and fire/EMS services. Fourth, an adequate financial base must be available for municipal, County, and fire/EMS services, and a plan to realign existing resources was designed. Fifth, an "end state" plan with defined boundaries was chosen rather than a more incremental or "process" approach. Finally, the board made specific proposals for two regional entities encompassing the Functional City: an Economic Development District to promote the area and a "Metropolitan Commission" that could be called into existence to deal with regional problems as they arise.

At closure, 37 new municipalities incorporating all of the County (see Figure 4.3) and four fire/EMS districts were proposed. A detailed fiscal profile for each new city was prepared to document an adequate balance between revenues and expenditures (Board of Freeholders, St. Louis City/County, 1988b). The proposed new cities ranged in population from 6,400 to 78,200, with only five having less than 10,000 residents. As a pivotal component, a fiscal plan realigned finances so that each government had a revenue base adequate to support public services.

Figure 4.3.   Map 3: Final Freeholders Proposal
SOURCE: St. Louis County Planning Department.

The comprehensive fiscal reorganization contained four elements. First was a new 1% countywide tax on earnings to fund County government and fire/EMS services. This would also fund a revenue-sharing pool for municipal resource equalization and infrastructure needs. Second was a new 6% gross receipts tax on nonresidential utilities to provide additional revenue for the 37 new cities. Third, all general sales taxes in the County would be distributed 25% point-of-sale, 75% per capita (pool). No longer would any city keep all sales taxes generated within its boundaries. Finally, property tax reliance would be reduced.

Several objectives were accomplished by this plan. First, all municipal services would be provided by a city, not the County. Second, resources would be balanced with needs, and future revenue growth would be enhanced through reliance on a sales and income tax base rather than property taxes. Third, each city and the County would be able fiscally to provide adequate basic services. Fourth, adequate fire/EMS protection would now be available countywide. Fifth, areawide issues would be addressed through two new regional entities. Overall, governance and fiscal structure would have been far better organized to deal with existing problems and to cope with any new regional issues that might arise.

■ Enter the Courts: The Legality of Governance Reorganization in St. Louis

During its work, the legality of the Board of Freeholders and its plan was challenged by a variety of citizens and public officials and as a result was reviewed by a variety of courts. The specter of comprehensive regional governance change was not popular, especially with local officials. Courts included the Federal District Court, the State Circuit Trial Court, the United States Court of Appeals for the Eighth Circuit, and the Missouri Supreme Court. Each court considered interpretation of "freeholder" as requiring property ownership for appointment to the board and thus as a denial of equal protection because it excluded renters.

After lengthy legal machinations, the board's legal status was affirmed by the Missouri Supreme Court on referral from the U.S. Court of Appeals. The battle, however, continued. On February 21, 1989, in a one-line order, the U.S. Supreme Court agreed to review the Missouri Supreme Court decision in its fall session, to begin after the already scheduled vote on the board's plan.

The board completed its work, all members signed the plan, and it was filed for submission to the voters. The board went out of existence on September 16, 1988. Voters would have exercised their prerogative on June 20, 1989, with a simple "yes" or "no" on the entire 155-page plan. The U.S. Supreme Court action delayed any vote pending its decision. For the Functional City, choice of a dramatically new governance and finance structure would have been an option available at the polls.

■ Resolution Absent Reform

The question of the constitutionality of the board had been granted *certiorari* by the U.S. Supreme Court. The case was argued on April 25,

1989 (*Quinn et al. v. Millsap et al.*), and a decision was delivered on June 25, 1989. The U.S. Supreme Court reversed the Missouri Supreme Court in a unanimous vote. It held that the land ownership requirement for selection to the board violated the equal protection clause of the U.S. Constitution.

This decision invalidated the Board of Freeholders and its plan, so no vote was held. Although the legal status of the board and its plan was settled, despite massive effort, nothing had been done to address the ongoing, fundamental issues confronting the St. Louis region—fragmentation, revenue disparities, haphazard boundary changes driven by fiscal self-interest, lack of areawide coordination, and the provision of adequate public services.

Given the U.S. Supreme Court decision, the option of being able to deal with regional governance issues through the Missouri Constitution was in doubt. The U.S. Supreme Court did not note changes, remedy, or removal in its decision. The Missouri Supreme Court heard arguments on January 3, 1990, as to whether this section of the constitution should remain intact, with "freeholder" to be changed to "qualified elector," or whether it should be deleted. This court decided to retain this provision, interpreting freeholder now to mean "qualified elector." As part of its decision, it mandated immediate appointment of a new board, a "Board of Electors," to prepare a new plan. With the legal status of the constitutional provision settled by the U.S. and Missouri Supreme Courts, a new board could revisit the plethora of governance issues that had been discussed over the past four decades and longer.

■ Governance Reform Again?: The 1990 Board of Electors Plan

A Board of Electors was appointed and began its work in July, 1990. During yearlong deliberations, it considered a wide range of plans and ideas, including true regional governance with the formation of a single county government including all the City and County—the Functional City (Board of Electors, St. Louis City-St. Louis County, 1991, p. 2).

It is interesting that this board did not make much direct use, if any, of the work by the 1987 board. By this time, the issue of comprehensive governance and/or fiscal reform had become political anathema. The objective seemed to be to fulfill the court mandate for a new board, to prepare a plan, and then to cease to exist, with minimal attention or controversy. It accomplished this!

The board's final plan had two proposals that were regional in scope. The first was to establish a Metropolitan Economic Development Commission to enhance employment opportunities within the Functional City (Board of Electors, St. Louis City-St. Louis County, 1991, p. 4). It would be financed by a 2% tax on nonresidential utility service. The second was to create a Metropolitan Park Commission to govern and improve all Commission parks (Board of Electors, St. Louis City-St. Louis County, 1991, p. 16). It would be funded by a tax not to exceed six cents per $100 assessed value on taxable property in the Functional City.

The Board of Electors plan (on the April, 1992, ballot) was defeated soundly. Given the litany of governance and fiscal issues facing the St. Louis area, it seems ironic that the board's plan avoided virtually all of them. Despite this, even its modest step toward regional governance was defeated; such change had become, or was still, unacceptable.

■ Enter Incremental Reform: A Boundary Commission

In the political aftermath of the Board of Freeholders process (but prior to the Board of Electors) and as a response to growing pressure to deal with the governance and fiscal issues in the state's most populous region, the Missouri legislature passed a bill to establish a St. Louis County Boundary Commission. According to its Mission Statement, the commission:

> is a review body which seeks to ensure that any municipal boundary change which takes place is in the best interests of the citizens of St. Louis County. The Commission's reviews and decisions are designed to enable all residents of St. Louis County to live within municipalities or unincorporated areas capable of providing adequate public services. (Boundary Commission, St. Louis County, 1994)

The commission began operation in 1990. It became clear that this approach to governmental and fiscal restructuring would be slow, incremental, and ad hoc in nature. Upon its review, the commission had authority to veto any annexation, consolidation, transfer of jurisdiction, or incorporation. An approved proposal could be placed on the ballot and must receive majority approval in each affected jurisdiction. As a review body, the commission did not have any authority, leverage, or incentives to encourage or mandate change. It was essentially reactive, not proactive, and operated in a contextual vacuum with no clear policy mandate to guide its activities. From the time it began its work, in 1990, through 1993, the commission received 60 proposals. Fifty-seven were annexations, two were incorporations, and one was a consolidation of two cities. Of these proposals, 17 were placed on the ballot, and 7 passed. All were annexations.

In April, 1993, the Missouri courts once again took action relating to governance/fiscal reform in the County. The Missouri Supreme Court found that the Boundary Commission legislation violated a clause in the Missouri constitution that prohibits "any special or local law regulating the affairs of cities and counties." The commission continued to operate, however, arguing that the offending clause already had been removed in 1992. All did not stop there. In May of 1994, the Missouri Supreme Court declared the County Boundary Commission to be unconstitutional and voided the entire statute under which it had been created.

The original 1987 statute clearly applied only to St. Louis County. Despite legislative revision in 1992, the Missouri Supreme Court held that the revision still was in violation of the constitution because it applied to St. Louis County alone. Because the offending provision could not be severed from the rest of the statute, the court ruled "the entire statute must fall."

■ Existing "Governance" Arrangements in the St. Louis Area

Governance arrangements and cooperation within the Functional City have been subject to the ebb and flow of politics and political personalities. Despite periodic conflicts, however, there is evidence of linkages over the past several decades. Examples include the Metropolitan Sewer District in 1954 and the Junior College District in 1962. In 1972, the Metropolitan Zoological Park and Museum District was created. It presently includes five key regional cultural institutions: the Zoo, Art Museum, Science Center, Botanical Garden, and History Museum. By 1984, cultural institutions and tourism also received support from the Regional Convention and Visitors Commission and the Regional Cultural and Performing Arts Development District. In 1985, the City Hospital and St.

Louis County Hospital were merged into St. Louis Regional Medical Center. More recently, in 1989, the St. Louis Regional Convention and Sports Complex Authority was established to provide Functional City and state financial support for a new domed stadium in the downtown area.

Although the Functional City does have a history of cooperation, a number of broader regional initiatives have been defeated. As noted above, in 1992 voters rejected a Board of Electors plan to create a Metropolitan Economic Development Commission and a Metropolitan Park Commission. These commissions would have provided a regional tax base for economic development and for the area's premier public park, Forest Park. Three earlier constitutionally board-based, regional governance plans have been defeated over the past 70 years (in 1926, 1955, and 1959).

This is not to suggest that the broader region does not cooperate in addressing policy issues: It does, but in a very limited way. In 1949, the St. Louis region, with an interstate compact between Missouri and Illinois, created the Bi-State Development Agency to address issues that cover two states and seven counties. Although its legislative powers and geographical scope are broad, its effectiveness has been limited by a dominant focus on transportation and a severe financial constraint in lacking any broad-based taxing authority except for a recently enacted regional transit tax for light rail. The Bi-State Development Agency might provide a vehicle for regional cooperation, but its limited financial powers prevent any proactive stance in dealing with regional governance. There is nothing on the horizon to suggest that this status will be altered in the near future.

In 1965, the East-West Gateway Coordinating Council was established as a not-for-profit corporation covering three counties in Illinois and five in Missouri. Its charge was broad, including transportation, regional planning, and metropolitan cooperation. It has a 30-year history in the St. Louis area, but its finances and authority are constrained. It continues to be very active and does planning, primarily transportation, for an eight-county, two-state area.

A more recent example of regional cooperation is the St. Louis Economic Adjustment and Diversification Program initiated in the late 1980s in response to dramatic cuts in the national military budget and the significant economic impact expected for the St. Louis region. In response, seven counties from both sides of the Mississippi River joined to initiate a program to assist laid-off workers and to help defense-dependent firms diversify into commercial markets.

Creation of the Bi-State Development Agency, East-West Gateway, and the St. Louis Economic Adjustment and Diversification Program had a broad regional base, albeit financially constrained. Such broad participation in a policy area is the exception rather than the rule. In most instances of regional cooperation, the coverage is limited to the Functional City and has taken the form of a new special district.

The number of policy issues in which the Functional City has been involved is unmatched by any other areas in the region. Although there are several policy issues in which the Functional City has participated, the nature of this cooperation has been unstructured, episodic, and limited. Often, there have been funding constraints as well, and cooperation has not included the other counties in the MSA.

■ The Present Status of Governance and Fiscal Reform in St. Louis

As of this writing, political/fiscal restructuring in St. Louis is at an absolute standstill. Indeed, it may be moving backward. The comprehensive reform proposed by the 1987 board was declared unconstitutional by the U.S. Supreme Court. Its 1990 successor board was appointed under court mandate, without which it certainly would not have been formed.

The 1990 board chose not to address fundamental issues of governance, fiscal disparities, and areawide services. Despite any merits, its proposals for economic development and park operation paled in contrast to the compelling issues facing the region. At this point the only remaining medium, as slow and cumbersome as it might have been, to deal with municipal boundary issues had been the Boundary Commission. This body also was declared unconstitutional. The full impact of this decision is potentially broad and fiscally devastating.

First, there is uncertainty about the proposals that have been received and acted on and those that are pending. At this point, one can only speculate. While legal questions remain unanswered, more than 3 years of work by this body is at risk. Absent this body, proposed boundary changes are not subject to review and are flourishing in a fiscally driven, land-grab environment; there is no "gatekeeper."

Second, the statutory language on which the Boundary Commission was declared unconstitutional relates to the exclusive identification of a single governmental jurisdiction, St. Louis County. This wording also affects a variety of other statutory provisions that relate to revenues within the County. There are very serious potential revenue loss implications for both the County and its municipalities if the Boundary Commission decision is extended to provisions for local revenues.

Although the extent of potential risk in not known at present, it could include a tax on general sales, gross receipts, motor vehicle licenses, cigarettes, and tourism and conventions. Coverage could reach in part or in whole to County government and to municipal governments. It should be noted that the statutory language of the type used for St. Louis County also raises a fiscal threat for numerous other political subdivisions in Missouri. It has been suggested that it could affect as much as one-third of all Missouri statutes.

■ What's Next for Reform in St. Louis?

Given the almost 70-year history of attempts to restructure governance and fiscal arrangements in the St. Louis area, one might ask "what's next?" The situation outlined above remains and may only worsen. In addition to County governance issues, there are concerns over broader regional problems such as zoning/land use, infrastructure, transportation, economic development, solid waste disposal, and sewage and waste water treatment.

The 1987 Board of Freeholders made provisions for an economic development district and outlined a Metropolitan Commission that could be called into existence to deal with areawide issues. The 1990 Board of Electors proposed a Metropolitan Economic Development Commission and a Metropolitan Park Commission. All of these would have operated on a Functional City basis had they come into existence.

Given the range and scope of considerations still to be dealt with, there is a strong case for calling into existence another Board of Electors. Although the past track record is dismal, there are few options available for reform. Of course, any proposal by such a board would still require voter approval, but it may be an opportune time to make another attempt for the following reasons. First, there seems little likelihood that the Missouri General Assembly can cope with all that is involved, given the scope and highly politicized nature of the environment. Any board-based proposal would go directly to the voters and would not require legislative authorization. Second, voter approval of a proposal satisfies the requirements of the so-called Hancock amendment to the Missouri constitution. This provision mandates almost literally (de facto politically) that *any change in any local revenue* must be voter approved. Since its statewide voter approval in 1981, the courts have adhered to a strict interpretation of its wording.

Third, the constitutional provision for a board allows it to deal with any issue affecting the Functional City. It also explicitly permits it to look into the operations of any local jurisdiction within this area, be it County, municipal, school district, or special district. Such a board can, de facto, come up with virtually any plan for any governmental/fiscal structure, for all or any part of the Functional City.

Fourth, the constitutional provision as it presently stands has undergone the scrutiny of every court with any jurisdiction, including the United States Supreme Court. Although the 1987 board was tainted, the 1990 board was not, nor would be any new board. The legal status finally seems unequivocal.

Focusing on a "process" rather than an "end-state" plan to accomplish the governance and fiscal objectives for the area might well be palatable to local and state political actors in getting their support and to the voters in getting their approval. Clearly, any such process would need teeth to accomplish objectives such as governmental reform and resource realignment.

Two other major issues that confront the area should be mentioned in closing. St. Louis school desegregation has been an ongoing legal and political issue at the local, state, and federal levels for the past three decades. It is worthy of consideration whether a new board could help move toward some resolution of this policy quagmire. The task would be Herculean, but there is much to be said for a local solution to a local problem, one that is presently dominated by federal courts. The other issue relates to the areawide Metropolitan St. Louis Sewer District (MSD). This entity is the one and only "success story" for the constitutional board provision discussed above. It was proposed and approved by voters in 1954. After functioning well for decades, presently it is immobilized by problems, most of which stem from finance. Its main sources of revenue are user fees and a property tax (Briggs, 1994).

At present, the MSD is unable to meet its expanding service demands plus state and federal clean water laws. It is in need of hundreds of millions of dollars for capital improvements, plus additional funds for operations. It cannot fund them, however, because courts have struck down past rate increases and the Hancock amendment requires voter approval of any revenue change; voters have refused. The MSD is in violation of both federal and state laws but cannot obtain funds necessary to bring it into compliance. Because the MSD was board created, perhaps a new board can rectify its current seemingly unsolvable dilemma before a federal court steps in and mandates change, which is not an unlikely scenario.

APPENDIX

■ Recent Events Affecting Regional Governance in the St. Louis Area

Since the original paper was prepared, numerous events have occurred that relate, either directly or indirectly, to the present status of regional governance in the St. Louis area. Following are the highlights of most of those events that are of significance. They are not listed in order of importance or impact.

Annexations and Incorporations

Two new cities have been incorporated in St. Louis County during 1995. Green Park has a population of about 2,400 in 1.3 square miles. Wildwood has a population of 17,000 in 67 square miles. It is larger in land area than the City of St. Louis, with only 5% of its population.

A present proposal would incorporate a new city called South Point that would have a population of 105,000 in 48 square miles. This area encompasses most of the unincorporated portion of south St. Louis County. There is at least one other active incorporation proposal in this same area.

The city of Bella Villa, with a population of about 700, recently proposed an annexation that would have increased its population to slightly under 10,000. It was defeated. Many other annexation proposals are under consideration or in the planning stage. During 1994, for example, 6 out of 14 annexation proposals received voter approval, affecting some 11,000 people.

St. Louis County Boundary Commission

The Boundary Commission that was declared unconstitutional because of its special legislation status (that is, affecting only one county) was re-created by the state legislature. This occurred after a statewide voter approval to amend the Missouri Constitution, allowing for provisions that relate to a single governmental jurisdiction. It has only just begun its work, but it will be, de facto, the "gatekeeper" for future boundary changes in St. Louis County.

Special Legislation on Local Revenues

The same special legislation issue that caused the Boundary Commission to be declared unconstitutional was also a serious potential threat to major sources of local revenues. In total, some 500 local laws were threatened; not all were revenue related. On April 4, 1995, a statewide vote approved a constitutional amendment that validated the so-called "special laws" that affected specific counties, thus removing this serious threat to financing local governments in Missouri.

Metropolitan Sewer District (MSD)

This jurisdiction, which provides water and sewage treatment for virtually all the St. Louis City and County area (some 524 square miles), has been plagued by a lack of funding for operations and capital because of a very restrictive constitutional limitation (the Hancock amendment) on local revenue increases. A recent study proposed turning the control of MSD over to a nonprofit organization. This would remove its rate increases from the requirement of voter approval under the 1980 Hancock amendment.

Consolidation of Fire Services in St. Louis County

A committee set up to study the St. Louis County Charter examined the possibility of setting up a single countywide fire and ambulance system. This would have consolidated the present 42 fire districts and municipal departments in the County. The proposal was defeated by a vote of 10 to 4.

St. Louis Regional Medical Center

This facility was established in 1985 by merging St. Louis City and St. Louis County hospitals into a single, jointly funded, regional health complex, the St. Louis Regional Medical Center. It served the area's needy residents for 10 years, but the County has indicated that it will not renew the contract for funding its portion. The situation is further exacerbated by changes in Medicaid funding, which seriously reduce funds going to the health complex. Its viability as a regional health facility is in question.

REFERENCES

Advisory Commission on Intergovernmental Relations. (1987). *The organization of local public economies.* Washington, DC: Author.

Advisory Commission on Intergovernmental Relations. (1988). *Metropolitan organization: The St. Louis case.* Washington, DC: Author.

Advisory Commission on Intergovernmental Relations. (1992). *Metropolitan organization: The Allegheny County case.* Washington, DC: Author.

Board of Electors, St. Louis City-St. Louis County. (1991). *Plan for Metropolitan Economic Development Commission and Metropolitan Park Commission of the City of St. Louis and St. Louis County.* St. Louis, MO: Author.

Board of Freeholders, St. Louis City/County. (1988a). *Plan for governmental reorganization in St. Louis & St. Louis County.* St. Louis, MO: Author.

Board of Freeholders, St. Louis City/County. (1988b). *Supplement to the plan for governmental reorganization in St. Louis & St. Louis County.* St. Louis, MO: Author.

Boundary Commission, St. Louis County. (1993). *1993 annual legislative report.* St. Louis, MO: Author.

Boundary Commission, St. Louis County. (1994). *St. Louis Boundary Commission information packet.* St. Louis, MO: Author.

Briggs, T. (1994). *An agency in crisis: The Metropolitan St. Louis Sewer District.* Unpublished memo.

Confluence St. Louis. (1987). *Too many governments?* St. Louis, MO: Author.

Metropolitan St. Louis Survey. (1957). *Background for action.* St. Louis, MO: Author.

Part II

Mutual Adjustment

5

Washington, D.C.:
Cautious and
Constrained Cooperation

JEFFREY HENIG
DAVID BRUNORI
MARK EBERT

The District of Columbia, in at least one sense, is the most isolated of cities. Like other large central cities, it is struggling to hold on to an economic and population base that increasingly is drawn to suburban locations. When taxpayers move out of the District, however, the fiscal body blow is unusually sharp and decisive. As the Federal City, the District of Columbia stands apart from the jurisdictions that surround it. Other cities have opportunities to soften the impact of suburbanization—using political leverage at the state level to push for taxing and spending policies in their favor or, perhaps, instituting a commuter tax to force at least some suburbanites to bear some of the cost of maintaining city services—that are not open to the District. The District's taxing and political power stops abruptly at its borders, and the congressional delegations from Maryland and Virginia work hard to maintain that status quo.

David Rusk (1993) has argued that healthy cities must expand or otherwise "capture" suburban growth. Out of 145 central cities with populations in excess of 100,000, Rusk scores DC as the sixth most "inelastic" (p. 132, Table A-1). His measure of inelasticity—which is based on population density and history of boundary expansion—understates the District's plight. Almost all the strategies he offers for stretching cities depend on action by a state government that incorporates all or most

AUTHORS' NOTE: This is a revised version of a paper originally presented at the 1994 annual meeting of the American Political Science Association, New York, September 1-4, 1994.

of the metropolitan area; such options are unavailable in the Washington case, where the metropolitan area spans DC, Maryland, and Virginia (Rusk, 1993, chap. 3).

The health and well-being of the District of Columbia, as the nation's capital, ought to be of general interest and concern, even if the particularities of its situation make it untypical in many senses. More important, the very peculiarities of the District's situation may bring into sharper relief some lessons that are relevant to other cities. As one of the most, if not the most, inelastic cities in the nation, the District provides an opportunity to assess whether Rusk overstates the degree to which inelasticity necessarily imposes severe consequences. In addition, the District's case brings to the fore the question of the federal government's role in promoting regional cooperation. Rusk makes it clear that, when it comes to stimulating movement toward the kinds of metropolitan government that he favors, "the 'action' is really with state and local governments" (p. 105). He identifies several ways that the federal government could encourage or require greater regional cooperation. The federal government's special interest in—and responsibility for—the District of Columbia makes this a likely setting in which the federal government might explore such options. The U.S. Constitution gives Congress exclusive jurisdiction over the Capital district, and—even after the passage of Home Rule legislation more than 20 years ago—Congress continues to play an active role in the governance of the city.[1]

The first section of this chapter summarizes some of the important demographic and economic characteristics of the Washington metropolitan region. Although the District government undeniably faces severe fiscal problems, the metropolitan area as a whole is doing rather well. Some sharp inequities exist between city and suburb, but it is not self-evident that this inequity has undermined the health of the rest of the region. The second section considers the status of regional cooperation and governance. Here, the focus is on the Metropolitan Washington Council of Governments (MWCOG), the premier multipurpose regional body in the area (Figure 5.1). Although the MWCOG has established a strong niche as a forum for intergovernmental communication and coordination, our discussion will highlight the political constraints that have led it to restrict its agenda to "safe" issues that do not even modestly challenge the governmental autonomy of the member jurisdictions. The third section presents an even more narrowly focused case study, analyzing the dynamics of participation of one MWCOG committee, the Police Chiefs' Com-

Figure 5.1. The Washington, D.C., Metropolitan Area Council of Governments Major Member Jurisdictions

mittee. Finally, the chapter offers some conclusions, focusing particularly on the impact of the federal government on regionalism.

■ Profile of the Washington Region

The Washington metropolitan area is undergoing something of an identity crisis. Until 1992, the area was defined by the Bureau of the

Census as comprising the District of Columbia plus 15 cities and counties in Maryland and Virginia. This is the area still designated as the Washington Metropolitan Statistical Area (MSA) and the area that will be referred to in most of the discussion in this section. In 1992, the area was expanded to include seven additional jurisdictions in Virginia as well as two counties in West Virginia. This larger area constitutes the Washington Primary Metropolitan Statistical Area (PMSA). In that same year, a new Washington-Baltimore Consolidated Metropolitan Statistical Area (CMSA) also was defined, linking the Washington area with the large Baltimore and much smaller Hagerstown, Maryland, PMSAs. The MSA's 1990 population of 3.9 million made it the seventh largest metropolitan area in the United States.

Rusk (1993) argues (following Savitch, Sanders, & Collins, 1992; Savitch, Collins, Sanders, & Markham, 1993; and Ledebur & Barnes, 1992) that suburbs surrounding inelastic cities will suffer, but the Washington metropolitan area is relatively vital and well off. From 1985 to 1990, the Washington metropolitan area grew at a rate of more than 84,000 people per year (Grier, 1993, p. 2), and its total growth of more than 672,000 for the decade ranked fifth among all metropolitan areas (U.S. Bureau of the Census, 1991, Table 2). Its 1988 per capita income of $23,175 was the highest in the country; its 1989 unemployment rate of 2.7% ranked 273rd (out of 281); and its per capita retail sales ranked 21st (U.S. Bureau of the Census, 1991).

The growth and well-being are not evenly distributed, however, and at least some disparities between the city and suburb are getting worse. Figure 5.2 illustrates population patterns and the racial composition of those shifts over the past three decades. Rapid growth in the suburbs is paired with a population loss in the District; in absolute numbers, growth in the large inner suburbs still outstrips that of the outer suburbs, although the latter are growing faster in proportion to their current populations. Between 1960 and 1990, the white population declined in the District, while it increased in the suburbs. Most of the city's loss of white residents occurred during the 1960s, accelerated, most likely, by court-induced changes in the educational system that led to greater equalization of spending, racial integration, and reduced reliance on educational tracking, as well as the major riot that took place in 1968. Indeed, between 1980 and 1990, the white population increased by a small amount. The suburbs, meanwhile, have been becoming increasingly diverse. Many black households have moved from DC into surrounding jurisdictions. Between 1980

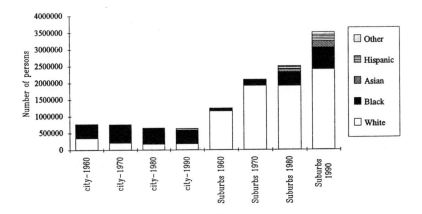

Figure 5.2. Racial Composition, the District of Columbia and Suburbs, 1960-1990

SOURCE: Data are from *Population Change in Metropolitan Washington: A Comparison of 1980 and 1990 Census Population and Housing Unit Counts by Subarea*, Department of Metropolitan Development and Information Resources (1991).

and 1990, the black population in the District decreased by more than 49,000 (11%); during the same period, the black population rose by nearly 122,000 (49.2%) in Prince George's County and 41,400 in Montgomery County (81.8%). Although the District boasts a small "Chinatown" neighborhood and a vibrant Hispanic commercial and residential life around the Adams Morgan and Mount Pleasant neighborhoods, by far the bulk of the region's rapidly expanding Asian and Hispanic populations is settling in the Maryland and Virginia suburbs. More than 88% of the region's net increase in Hispanics between 1980 and 1990 and more than 95% of its net increase in Asians took place outside the central city (Metropolitan Washington Council of Governments [MWCOG], 1991, Tables 5 and 6).

The large and growing minority populations in the suburbs mute the sense of racial polarization somewhat, but the disjunction remains a sharp one between the majority black District and the majority white suburbs. This racial-demographic disjunction is overlaid by a more sharply etched racial-political cleavage: Whereas the District's public officials are predominantly black, the growing minority population in the suburbs has not yet been translated into proportional representation among the political leadership (Gale, 1987, chap. 11; Henig & Gale, 1987). Sharper than the racial cleavages is the economic disparity between city and suburbs. Although the District boasts a relatively high median income, when compared to other large cities, the ratio of the median income of District households to that of suburban households has been declining sharply (see Figure 5.3).

In spite of its extreme inelasticity, however, the District is by no means among the worst off of large metropolitan areas, even in terms of some of the basic indicators of regional equality. For example, the District ranks 12th, out of the 39 largest metropolitan areas, according to the extent to which the city has been forced to absorb a disproportionate share of the metropolitan area's neediest residents (calculated from Grier, 1993, p. 26). Harrison and Weinberg (1992) found DC to rank in the middle among large metropolitan areas in segregation for Blacks (19th out of 44), Hispanics (22nd), and Asians (20th) (Tables 4, 8, and 12). Of the 30 MSAs (regardless of size) with more than 25% Blacks, the Washington MSA ranked 26th in black segregation (Harrison & Weinberg, 1992, Table 5).

■ The MWCOG: Regional Cooperation, Within Limits

The MWCOG is the premier intergovernmental organization in the Washington area. It was founded in 1957, well before the federal government required metropolitan areas to form regional councils to qualify for federal grants in the mid-1960s. Indeed, only three similar organizations were in existence at the time the MWCOG was established.[2]

Since its founding, the MWCOG has grown from seven members with no staff and an annual budget of less than $20,000 into a $10 million organization comprising 18 jurisdictions, 120 employees, and spacious offices located near the Capitol. Having begun with a modest agenda of lobbying Congress and state legislators, the MWCOG today is involved in myriad complex issues ranging from transportation planning to environmental cleanup efforts.

Despite nearly four decades of sustained growth, the MWCOG has remained a fragile coalition of governments representing diverse political and economic interests. It has stayed together, indeed prospered, because it has carefully avoided controversial issues while providing its members with tangible benefits. This section discusses how MWCOG maintains this delicate balance.

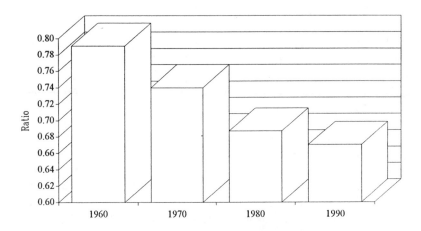

Figure 5.3. Ratio of Central City Median Household Income to MSA
Median Household Income, 1960-1990
SOURCE: Data are from *Census of Population and Housing*, U.S. Bureau of the Census, various years.

Avoiding Controversy—Structural
and Political Limitations

A review of the MWCOG's history reveals an organization unable to
address controversial issues. This caution is rooted in a structural and
political development that resulted in limitations that exist to this day.
Structurally, the MWCOG has no governmental authority. It has no
revenue-raising ability and is completely dependent on federal and state
grants (60% of revenue), contractual fees (30%), and member contributions (10%). The MWCOG has no police or regulatory powers and no
legal means of forcing its members to take any action. These are not minor
limitations, for they effectively preclude the MWCOG from taking on the
most serious problems facing the region.

The MWCOG is a nonprofit, voluntary association of county and
municipal jurisdictions. No jurisdiction is obligated to join or to remain
a member. Any member can withdraw at any time and for any reason. It
is not an accident that the MWCOG is a voluntary organization. Efforts
in the early 1960s to obtain formal governmental status through either a
Congressional Charter or an interstate compact failed because of fears by
some members that the MWCOG would gain too much power over local
governments (see Scheiber, 1992, pp. 9-10). Because it is voluntary, the

MWCOG must consider carefully the impact of its actions on each of its members.

The obvious consideration in this regard is that members will leave the MWCOG. Indeed, members have left or threatened to leave the MWCOG for various reasons over the years. For example, Montgomery County withdrew in 1963 after conservative Republicans gained control of the county council. "Several members of the County Council viewed it as too left-wing in orientation, while others spoke of it as the possible precursor of a metropolitan government for the Washington Area, something they feared could erode their powers as local officials" (Scheiber, 1992, p. 6). There was great concern within the MWCOG that other jurisdictions would follow suit. Montgomery County would later return, but only after the MWCOG restructured its governance to give larger jurisdictions a much stronger formal role than those with populations under 100,000.

The state of Virginia, similarly, almost torpedoed the organization before extracting concessions that weakened the MWCOG's formal authority. In 1967, the state's attorney general ruled that Virginia law prohibited local governments from making financial contributions to any organization with headquarters outside the state's boundaries (Scheiber, 1992, p. 16). This would have forced all seven members from Northern Virginia to drop out. Fortunately for the MWCOG, intervention by a supportive state legislator led the attorney general to soften his original position: Local governments could appropriate funds for the MWCOG, but only if those locals maintained "some supervision" of MWCOG activities that would directly affect Northern Virginia. More recently, a combination of budgetary reasons and a sense that their interests differed from those of the more urban jurisdictions closer to the region's core led Charles County and the City of Frederick to withdraw in 1988. The City of Frederick returned to the MWCOG in 1994.

Coupled with the structural limitation of being a voluntary organization, the MWCOG has long faced the reality of existing in an area divided politically, economically, and racially. These divisions lessen the MWCOG's effectiveness in dealing with the region's more politically volatile public policy issues.

It is not surprising that many of the political divisions involve issues pitting DC against its suburban neighbors. For example, in 1968, the MWCOG, at the suggestion of DC, proposed a strict model gun control law for consideration by the members. The proposal caused considerable public uproar, especially in suburban Maryland, and the effort was abandoned. Realizing the sensitivity of the issue, the MWCOG refused to join DC in advocating regionwide gun control legislation in 1975.

More recently, the MWCOG repeatedly has declined to consider plans that would allow DC to levy a commuter tax. The suburbs upon whose citizens the tax would fall are strongly and unanimously opposed to such a tax. Similarly, the MWCOG refuses to endorse calls for statehood for DC, another issue soundly opposed by the suburban members.

The MWCOG's avoidance of volatile issues is not limited to disputes between DC and the suburbs. Rather, the MWCOG must avoid acting in areas that might be perceived as favoring a specific jurisdiction. Economic development is one such area. The MWCOG abstains from initiating economic development projects that would benefit discrete member jurisdictions. The MWCOG's avoidance of such economic development activities is not the result of reluctance on the part of its professional staff, which generally supports such endeavors; rather, it is a function of the inherent structural limitations of the organization. Although the members profess shared interest in the economic health of the region, their allegiances are more typically shaped by a narrower jurisdictional self-interest. The MWCOG jurisdictions all strive to obtain those development projects that will result in additional tax revenue; indeed, the various jurisdictions often compete vigorously for businesses that wish to relocate into the metropolitan area.

In addition to the political/structural limitations that deter activity in the area of specific economic development initiatives, the MWCOG is hampered by limited finances. The MWCOG receives no federal or state funds relating expressly to economic policy issues. Thus, activities concerning economic development would have to be financed through member contributions. Hesitation about spending limited resources on such activities has contributed to the MWCOG's relatively passive role even in pursuing broadly framed regional economic development initiatives. For example, the Greater Washington Board of Trade proposed in 1993 that the COG develop a regional marketing strategy aimed at attracting businesses to specific locations in the metropolitan area. Although the MWCOG supported the goal of developing such a regional strategy, it declined to fund the initiative.

Instead of making policy recommendations as to where development should occur, the MWCOG is limited to conducting regional studies of growth through its Cooperative Forecasting Program. These studies focus on residential growth, transportation needs, and sewage/water services.

This general regional information is used by local economic development offices to lure prospective businesses. Some MWCOG activities, particularly in the transportation area, have clear economic development consequences—after all, better schools, roads, and infrastructure tend to attract business investment—but the MWCOG refrains from taking on location-specific economic development projects.

Just as the MWCOG avoids economic development issues because it chooses not to be seen as advocating policies that will benefit specific jurisdictions, the organization generally avoids taking on issues that would result in a perceived detriment to a particular member. Housing issues are a primary example. The MWCOG once distributed millions of dollars of federal housing assistance to its members. When the federal government dramatically cut housing grants and changed the method through which its grants were distributed,[3] the MWCOG's involvement in this area also declined. Although the MWCOG has endorsed the broad principle of regional fair-share housing, it has been wary of recommending strong and specific measures that would directly require some member jurisdictions to increase their low-income housing. Without the leverage provided by strong federal support, the MWCOG has been relegated to minor actions that support, rather than challenge, member jurisdictions' self-defined housing goals.[4]

The MWCOG's lack of governmental powers and its desire to avoid controversy have allowed the MWCOG to stay together. The cost, however, is that these limitations have attenuated its effectiveness in addressing important issues commonly faced by urban governments.

The Benefits of Regional Cooperation

Despite the limitations outlined above, the MWCOG has survived because it provides tangible benefits to its members. Indeed, many of the functions carried out by the MWCOG would never be as successful, or even possible, if attempted by an individual local government. The MWCOG performs two primary functions: It distributes federal and state funds to its members, and it provides contractual services in the form of public policy research, information dissemination, and bulk purchasing.

Distributing Federal and State Funds. Distributing federal and state funds to its members has been the MWCOG's most important activity since the 1960s. Federal law has long required transportation, housing, and environmental grants to be distributed through regional organizations. Those jurisdictions that are not part of such organizations generally are ineligible to receive federal funding.

Currently, the MWCOG's main distribution effort is to direct the use of approximately $250 million each year for highway and transit improvements under the federal Intermodal Surface Transportation Efficiency Act (ISTEA). ISTEA provides federal funds for road repairs and infrastructure improvement, as well as for planning to reduce congestion, to meet air quality requirements, and to promote desirable development patterns. Like many federal programs before it, ISTEA requires that funds be distributed through metropolitan councils.

Under ISTEA, each local jurisdiction identifies repairs or other projects for which it would like funds and submits requests to the MWCOG. The MWCOG evaluates each request and, based on its evaluation, allocates the federal funds to the member jurisdictions. The MWCOG also monitors compliance with ISTEA rules.

The MWCOG also distributes federal and state funds for environmental protection activities in the region. Since the early 1960s, the MWCOG has devoted considerable time and effort, through its Water Quality Management Committee, to fund cleanup efforts along the Potomac River. According to MWCOG data, pollutants in the river have been reduced by 90%. Other important environmental activities funded through and coordinated by the MWCOG include regional recycling programs; establishment of the Small Habitat Improvement Program to clean up local marshland; a 1993 study of air quality in the region, concluding that ozone pollution was a major threat; establishment of a system for removing hazardous wastes from local businesses; and cleanup efforts along the Anacostia River.

The MWCOG unquestionably provides benefits to its members in performing its distribution functions. In most cases, the federal and state funds can be distributed only through regional organizations. Thus, belonging to the MWCOG is requisite to receiving intergovernmental aid. Moreover, problems such as traffic congestion and pollution are regionwide. They cannot be addressed by a single jurisdiction.

Interlocal Services. In addition to distributing and monitoring federal funds, the MWCOG provides substantial interlocal services to its members on a contractual basis. For example, the MWCOG, through its staff of transportation engineers, collects and analyzes data on traffic patterns, roads, and highways. Local jurisdictions use the data to determine repair needs, the placement of traffic signals, the need for additional infrastructure, and zoning regulations.

In addition to providing direct services to its members, the MWCOG provides public policy research for and disseminates information to local governments. The MWCOG conducts extensive research on a wide variety of topics such as alternative transportation systems, responses to the AIDS crisis, prison overcrowding, teenage drinking, and child care services. The MWCOG formulates policy alternatives and often coordinates regional responses to these and other problems.

For example, after the Air Florida and Metrorail disasters in 1981, the MWCOG developed a plan for handling emergencies across jurisdictional lines. The MWCOG coordinated the establishment of an emergency communications system and joint training for area fire and police departments. Similarly, in response to several major snowstorms in the 1980s, the MWCOG developed a snow emergency plan that has been used extensively by MWCOG members. Through the MWCOG, the federal, state, and local governments adopted snow priority routes, a staggered work release schedule, and a communications/computer system linking emergency health care providers.

Perhaps the most tangible benefit that the MWCOG provides to its members is the ability to make purchases through the Cooperative Purchasing Program. Under this program, now entering its 11th year, member governments buy oil, gas, and other supplies and equipment through the MWCOG. The MWCOG purchases the products in bulk and passes the savings on to its members. The MWCOG estimates that the program has saved its members more than $1.8 million a year and more than $20 million since the program's inception. For smaller members, the program has saved far more than their contributions to the MWCOG. More recently, the MWCOG has developed the Health Care Coalition, designed to allow members to purchase health care insurance at reduced premiums.

■ Law Enforcement:
A Case Study in Cooperation

That the MWCOG does not measure up as an incipient regional government will relieve some and disappoint others, but it should not surprise anyone; governance, after all, is not what the MWCOG was designed to provide. Although the MWCOG primarily may be a reactive organization, unable to set its own agenda and forced to tiptoe gingerly around issues of consequence, its ability not only to survive but also to grow over the last 37 years makes it apparent that it has carved out its own niche in the region. What accounts for the MWCOG's survival? Given the fact that member jurisdictions are free to leave the organization at any time, their allegiance suggests that they perceive distinct benefits. Are they held in the group by a unifying commitment to a shared metropolitan identity? Do they maintain ties simply as a way of keeping their eye on potential regional competitors? Or are there selective benefits that accrue to those who share information and find ways to cooperate with their neighbors?

This section seeks some insight into the inner workings of the MWCOG in order to gain some appreciation of the glue that holds the group together. Law enforcement is far from the largest component of the MWCOG's budget, and the Police Chiefs' Committee is not among the largest, most active, most influential, or most controversial MWCOG units. Nor, for that matter, can we make any claim that it is somehow the most "typical" of the organization's enterprises. In the light of the MWCOG's history—which suggests that public and controversial issues are more likely to sow seeds of division than to provide a rationale for cohesion—it makes some sense to look in such a quiet corner for insight into the dynamics of cooperation. Moreover, law enforcement presents enough of a dilemma to regionalism to make things interesting. Because criminals are very free to move across jurisdictional boundaries, law enforcement is an area that seems highly appropriate for metropolitan cooperation. Given the intense public concern about crime and a tendency by suburbanites to link crime with central cities, however, it also is an area potentially open to divisiveness and interjurisdictional finger pointing.

The Police Chiefs' Committee

Designed primarily as a vehicle for the sharing of information, the MWCOG Police Chiefs' Committee usually meets seven times per year at the MWCOG office in Washington, D.C. The top law enforcement official of each of the 16 member jurisdictions that have police departments is afforded committee membership with full voting rights. Activities of the Police Chiefs' Committee include review and development of mutual aid operations plans, sponsorship of police training seminars,

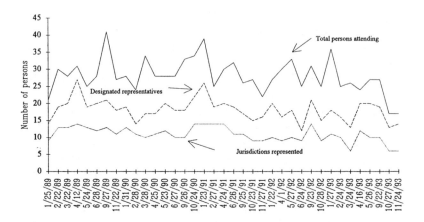

Figure 5.4. Level of Participation in the Police Chiefs' Committee, 1989-1993

SOURCE: Attendance sign-up sheets, MWCOG Police Chiefs' Committee.

exchange of information about crime trends, operation of five traffic programs, discussion of hot-pursuit guidelines, preparation of reports on area crime statistics, and cooperative purchasing of new technology.

Patterns of Participation

Attendance sheets for the 37 committee meetings from 1989 through 1993 were coded in an effort to measure the extent and levels of participation by the various member jurisdictions.[5] Our coding scheme was designed to measure attendance levels, indicated by the presence of *any* representative, and commitment levels, indicated by the relative rank of the representatives sent.

Figure 5.4 illustrates attendance data by meeting date. The three lines indicate the total number of persons attending, numbers of those attending as chiefs or official representatives of chiefs, and the total number of jurisdictions with at least one official representative present. All three lines tell a similar story. First, attendance is fairly good: No meeting had fewer than 17 persons attending. Second, there is no evidence of a sharp trend in participation. There is a mild decline in attendance over time, especially when measured by number of jurisdictions with a representative present, but this is largely accounted for by the last two meetings.

In the light of the general instability of the pattern, this may turn out to be an idiosyncratic and short-term drop. Third, attendance spikes occasionally, suggesting that some meetings with more interesting agendas may have drawn more people.

Do these highly attended meetings reflect agendas laced with controversial issues or decisions with the potential to redirect, in any substantial way, the policies or priorities of the member jurisdictions? A review of the agendas and minutes for the three best-attended meetings suggests that this is not the case. Agenda items for the September 27, 1989, meeting included collection and distribution of hate violence incident reporting data, addition of sexual orientation and handicapping condition to the MWCOG Hate/Violence Guidelines, and restriction of fuel deliveries to nonpeak traffic hours to reduce danger on the Capital Beltway. Among the agenda items for the January 23, 1991, meeting was the status of the license application for a proposed new television channel on a frequency that could potentially interfere with police broadcasts. The January 27, 1993, meeting agenda included introduction of newly installed Metropolitan (Washington, D.C.) Police Department Chief Fred Thomas and a briefing by the FBI (later tabled until February) on the Metro Area Violent Vehicle Theft Task Force computer system. Although some of these issues might have important components, at least on their face they do not appear to be substantially more controversial or significant than issues covered at the less well-attended meetings. For example, the items on the agendas of the meetings on January 25, 1989; October 27, 1993; and November 24, 1993—the three meetings with the lowest attendance over the 5 years —included a status report on the advisability of a regional inmate work release policy, approval of a proposed resolution on cameras at intersections, elections of committee chair and vice chair, and action on proposed new "hot-pursuit policy" guidelines.

If not broad substantive import, what is it about the agendas that might account for the higher attendance? One possible link among the agendas of the heavily attended meetings is attention to matters with the potential to affect law enforcement agencies' workloads. Together, the two best-attended meetings had four agenda items that conceivably would have required law enforcement agencies to assign additional responsibilities to their officers, possibly distracting them from other activities that the chiefs consider more important. A second possible explanation is the added interest that may accompany meetings at which there are guest speakers (e.g., the introduction of DC's new police chief at the January 27, 1993, meeting) and nominations for officer of the year awards. A third

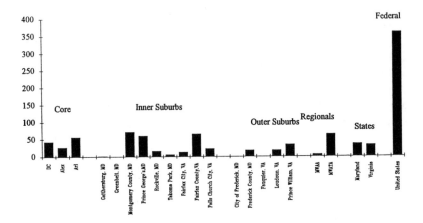

Figure 5.5. Total Numbers of Persons Attending, 1989-1993, by Jurisdiction (grouped by zone)
SOURCE: Attendance sign-up sheets, MWCOG Police Chiefs' Committee.

possibility is that participation levels may reflect the amount of time that has elapsed since the prior meeting. Each of the three most highly attended meetings occurred after 3 months had passed without a committee meeting. Such a pattern would make sense, for example, if one of the benefits members see in the meetings is the chance to exchange information occasionally and refresh personal ties with counterparts from other jurisdictions. Those with looser ties to the group might be less likely to attend, accordingly, when meetings quickly follow on the heels of one another.

Figure 5.5 illustrates the high degree of involvement by representatives from the federal level of government. Over the 5-year period, 361 people from more than 20 federal agencies attended the meetings—243 as designated representatives and the others as observers, guests, or subcommittee chairs.[6] Core jurisdictions and the inner suburbs, on average, send more participants than outer suburbs. Thus, there is some indication that, as a group, jurisdictions closer to the central city have greater participation.

In addition to proximity, the size and bureaucratic complexity of the jurisdictions may relate to participation as measured by meeting attendance. Particularly at the local level, larger jurisdictions send more people than do smaller ones (for example, counties send more representatives than municipalities). This largely may be a case of limited human resources: Some of the smaller jurisdictions may not be able to spare people to attend such meetings in the normal course of events.

Analysis of the ranks of those attending shows that, generally, representatives are highly placed within their agencies—typically the chief or a top-level substitute. Lower-rank representatives are most likely to come when a higher-rank official also is attending. Although outer suburban jurisdictions send fewer participants, those they do send tend to be more highly placed within their units. The relatively steady involvement by fairly senior officials suggests that they find these committee meetings worthwhile. Their attendance certainly is not attributable to broad public pressure for involvement; these meetings are not covered by the media, and it is safe to say that most area residents have no idea that they are taking place. Nor does it appear that attendance is defensive in nature, as might be the case if chiefs were concerned that action taken in their absence might infringe on their territory or deal them out of a rewarding enterprise. The formal agendas, at least, give no indication that such issues are at stake. Rather, it would appear that representatives are responding to the opportunity to share information, to maintain contact with colleagues facing some similar challenges, and to engage in some noncontroversial joint undertakings for which there are efficiencies of scale.

The very strong involvement by the national government undoubtedly is due in large measure to the special characteristics of the DC metropolitan area. Although many other metropolitan areas also have local offices of federal agencies, certainly none boasts as many potential federal-level partners. This heavy national-level involvement might provide the DC area COG with some special advantages in terms of access to inside information, for example, or even an inside track on some competitive categorical grants. It is also worth considering the general implications for our understanding of metropolitan organizations. Rather than straightforward collections of local jurisdictions, regional bodies like the MWCOG may comprise nested arrangements of governments of different levels.

■ The Impact of the Federal Government on Regionalism in Its Own Backyard

David Rusk, an unabashed advocate of regional approaches, nevertheless sees the federal government as having only a limited role to play in their encouragement. His wariness toward federal solutions rests on his reading of the U.S. Constitution, which leaves sovereignty in matters of local governmental structure to the states (Rusk, 1993, p. 102), and on his preferences for decentralized solutions rooted in his personal experience as a mayor and state legislator. In this section, we consider the role that the federal government has played in shaping (and failing to shape) regional institutions for governance in the Washington metropolitan area.

Because of the national government's special stake in the governance and well-being of the District, it seems reasonable to look here for evidence that the federal government has the will and capacity to play a more forceful role than Rusk suggests. In some ways, the federal government's special interest in the District has contributed to regionalism; in some ways, it has worked against it. Overall, we conclude that the federal government has had a tremendous impact on regionalism in the Washington metropolitan area but that this impact for the most part is *not* attributable to the area's special status as home of the national government. Politics and philosophies operating at the level of the federal government, we suggest, account for the broad ebbs and flows of regional solutions throughout the country. That the growth, style, and agenda of the MWCOG reflect national policies is indicative of general patterns, not local peculiarities, leads us to some general conclusions about the prospects for regional governance in the years ahead.

DC's Special Status: A Two-Edged Sword

It would be reasonable to expect some additional support for regionalism because of the special interest of the federal government. For several reasons, national officials have sometimes looked to the District as an appropriate site for demonstrating broad new policy initiatives. One reason is the sense that DC is a natural "showcase" and therefore a tactically desirable site for sending a message to the nation. When Lyndon B. Johnson launched his program for "new towns in-town," for example, he especially wanted to get off to a quick start in DC to "give the program momentum and show a skeptical public that the government really meant to act" (Derthick, 1972, p. 25). More recently, Secretary of Housing and Urban Development Henry Cisneros and Deputy Secretary of Education Madeleine Kunin each have made it a point to focus special efforts on the District. Cisneros launched a special initiative to deal with homelessness in the District, and Kunin established a special "DC Desk" to monitor DC education policy and to seek ways in which the department could facilitate local efforts at school reform. A second reason for focusing national attention on the District is that the federal government's oversight power makes it potentially easier for Congress and the White House to manipulate the local agenda. A third reason is that many national officials live in the city or its suburbs; this can make them especially alert to the region's problems and gives them a more direct and personal stake when traffic congestion makes the commute from the suburbs too time-consuming, for example, or when crime in the ghetto neighborhoods spills over onto Capitol Hill.

The history of regionalism and the MWCOG provides some examples of this special status. As we have seen, the fate of regionalism in the Washington area has been tied closely to the federal government from the very beginning. Although the formation of the MWCOG slightly predates the concerted federal effort to encourage regional planning throughout the 1960s and 1970s, the growth of the organization, its choice of policies on which to focus, and its general mode of operation have reflected federal policies in many important respects. Although this is true of regional councils generally (see below), there are at least some senses in which the presence of the nation's capital has exacerbated those effects. Indeed, during the early years of the MWCOG's development, Congress was directly responsible for funding the enterprise. When the MWCOG drew up its first budget in 1958, the members initially envisioned that Congress— as the overseer of pre-Home Rule DC—would fund the entire $19,000; when Congress appropriated only $13,000, the other member jurisdictions agreed to take up the slack by contributing $1,000 apiece (Scheiber, 1992, p. 4). As our analysis of attendance at the Police Chiefs' Committee indicates, federal officials have been regular participants in MWCOG enterprises; there seems little doubt that this has facilitated the flow of information across the levels of government. Lacking comparable information from other COGs around the country, we cannot say for certain that this federal involvement is substantially greater than elsewhere, but it seems likely that it is.

The federal role, however, has been a two-edged sword. The unique governance arrangement imposed on the District of Columbia by virtue of its role as the home of the national capital also serves as an impediment to regional solutions. Most particularly, the ability of congressional members from Maryland and Virginia to constrain the policy prerogatives of the District's elected leadership has sharpened the city's resentment of its suburban neighbors. That this resentment built on unequal power has a racial component as well makes for a highly charged environment in which overtures toward regionalism are treated gingerly.

During most of the twentieth century, the key committees in the Senate and House were chaired by Southern segregationists overtly hostile to the city's black residents. The chairman of the House Subcommittee on District Appropriations in the late 1930s cut expenditures for welfare and education, observing that "My constituents wouldn't stand for spending money on niggers" (Jaffe & Sherwood, 1994, p. 27). The chair of the Senate District Committee from 1943 to 1947 was a member of the Ku Klux Klan who wrote a book called *Take Your Choice—Segregation or Mongrelization* (Jaffe & Sherwood, 1994). During the 1950s and 1960s, Congressman John McMillan from South Carolina generally was regarded as the de facto "mayor" of DC, and he systematically used his power to thwart efforts to extend the city's power to govern its own affairs.

When Congress took initiatives with a regional thrust, as often as not it was at the behest of national lobbies or suburban jurisdictions, sometimes directly in opposition to expressed preferences of DC citizens. In 1968, 95% of District citizens voting in the Democratic primary approved a proposal that all major transportation projects be submitted to a local referendum, but this proposal had no official standing, and Congress chose to ignore it. The next year, an unofficial citizens' referendum found that 84% rejected construction of a planned bridge and associated freeways, but Congress continued to insist on these (threatening to hold up all federal payments to the city until the project was begun). As a local advocate for stronger Home Rule observed, "A congressman may vote for construction of the Three Sisters Bridge . . . and thus pay a debt to the highway lobby; but he would not dare vote for such a project in his own district in the face of such opposition" (Smith, 1974, p. 149).

Although Congress has played a less intrusive and more supportive role over the past two decades, the limitations on self-government maintained under the current Home Rule legislation continue to rankle. One of the most controversial foci of tension directly relates to regional cooperation. Under the terms of the District's Home Rule authority, the city is prohibited from instituting a tax on the wages of suburban residents who commute into the city to work. City budget officials estimate that such a tax would have generated more than $1.2 billion in revenues in 1992 (Office of Policy and Evaluation, 1992, p. 114). The congressional delegations from Maryland and Virginia are adamant in their opposition to such a tax, and their colleagues have shown no inclination at all to challenge them in this regard. Moreover, the threat of suburban resistance presents severe political obstacles to even much subtler and more indirect efforts by the District to recoup some of the costs associated with serving a broad metropolitan population. When Mayor Sharon Pratt Kelly proposed, in early 1994, a sharp increase in the city's parking tax, opponents tactically painted the measure as a hidden tax on suburbanites. "This is not a parking tax," a spokesman for the association representing private owners of parking lots declared, "this is a commuter tax" (Henderson, 1994). When it became clear that the Council was not going to back her on this issue, the mayor was forced to back down.

In this context, it may not be surprising that many District residents and officials fear that regional initiatives could present one more arena in which suburban jurisdictions are able to erode the city's power to govern itself. What is more, the concerns based on the District government's formal dependency on Congress are made much more potent by the underlying belief, shared among many of the city's black residents, that there is a white conspiracy to recapture the city. This conspiracy, referred to colloquially as "The Plan," is believed to link Congress, the suburbs, and the business community in a broad plot to repopulate desirable neighborhoods through gentrification and to regain political control through federal intervention (Pianin & Milloy, 1985). It is not clear whether the District's political leaders harbor such fears themselves, but it *is* clear that appeals to such fears have the potential to spark powerful grassroots sentiments. DC politicians find themselves playing to those sentiments, sometimes, as a means of rallying support (much like suburban politicians sometimes raise the specter of central city problems rushing across the border to rally support for themselves). Although not at all unique to the Washington metropolitan area, such an intertwining of regionalism with racial issues may be exacerbated there because of its peculiar governance institutions.

Reflecting National Priorities and Trends

In arguing that DC's special status has had mixed effects on the development of regionalism in the Washington area, we are not suggesting that the aggregate federal impact has been minor. To the contrary: The ebb and flow of the MWCOG's influence has reflected broad national policy directions. Federal initiatives made it possible for the MWCOG to build its budget and staff, to extend the range of decision-making processes in which it played a legitimate role, and to provide selective incentives to member jurisdictions that made their continued participation worthwhile. In these respects, Washington is not much different from the rest of the country. Rather than emerging from localized recognition of shared metropolitan interests, the bulk of the nation's experimentation with regional government appears to us to have been stimulated and underwritten by national policies.

The most dramatic federal initiatives to encourage regionalism came in the 1960s, although the momentum carried through much of the following decade (see Atkins & Wilson-Gentry, 1992). The first federal program providing for intergovernmental coordination of planning was contained in a 1959 amendment to the Housing Act of 1954. As implementation took hold in the early 1960s, funds available to local jurisdictions rapidly increased (Atkins & Wilson-Gentry, 1992, p. 468). Direct eligibility was extended to regional councils of governments in Section 701(g) of the Housing and Urban Development Act of 1965, which provided that "regional councils of elected officials" would be eligible for federal assistance for regional planning on a two for one matching basis. Before 1966, there were ap- proximately 175 regional councils nationwide; over the next 4 years, this number tripled (Atkins & Wilson-Gentry, 1992, Table 1). In DC, Section 701 gave the MWCOG a big edge over the National Capital Regional Planning Council (NCRPC), a federal agency with which it had been jockeying for a preeminent planning role (Scheiber, 1992, pp. 8-9). The MWCOG, as a body compris- ing elected officials, met the eligibility requirements, whereas the NCRPC did not.

The Model Cities program and the Metropolitan Development Act of 1966 also extended areawide grant review powers to areawide planning agencies. The reach of these initiatives was extended by the Intergovernmental Cooperation Act of 1968. Circular A-95, issued by the Office of Management and Budget (OMB), outlined the procedures through which the act's provisions were to be implemented. Regional councils were given a key role in reviewing and commenting on local governments' applications for a wide range of federal categorical grants. According to Atkins and Wilson-Gentry (1992, p. 469), at the time the regulations were put into effect, 259 federal programs had A-95 review and comment requirements. In DC in 1968, federal money was flowing to the MWCOG "in unprecedented amounts" (Scheiber, 1992, p. 19).

Figure 5.6 shows that the dramatic growth of the MWCOG budget beginning around 1967 was largely underwritten by federal grants. During the first 3 years for which this budget data is available, most of the MWCOG budget was derived from local dues. Beginning in 1966, the

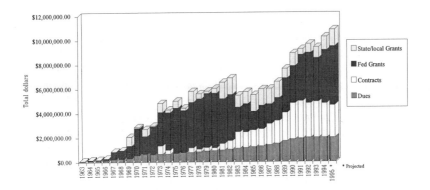

Figure 5.6. Revenue, by Source, 1963-1995
SOURCE: *Annual Report*, Metropolitan Washington Council of Governments, various years.

federal contribution consistently has outstripped member dues, with the ratio running about 4.5:1 during the 1974-1980 period.

To some extent, the federal initiatives were designed to encourage local areas to articulate their own regional plans, shaped by local values and visions. Through the carrot more than through the stick, the federal government also was seeking to alter the substantive agendas that these regional bodies would take on. In areas such as transportation, that model was akin to "cooperative federalism" (Grodzins, 1966; Henig, 1985); local and national goals did not vary greatly, and the federal support was welcomed. In other areas—such as those involving the environment and fair housing—the federal government's push toward regional solutions was seen as more problematic by local officials concerned with preserving local prerogatives. The MWCOG's efforts to take a stronger role in overseeing the region's compliance with amendments to the federal Water Pollution Control Act, for example, in 1974 led "an irate James Gleason, Montgomery County Executive, to the COG table for a series of skirmishes, to threaten Montgomery's withdrawal from the Council if a new unit came into being and to warn that COG was beginning to encroach on local prerogatives" (Scheiber, 1992, p. 26).

By the mid-1970s, however, the federal government's support for regionalism was beginning to level off, and federal cutbacks were no less influential in affecting the course of regionalism in the DC area as had been the years of expansion and largesse. According to then-Director of the MWCOG, Walter Scheiber, 1968 "probably was the last year in which the future of regional cooperation seemed almost limitless. From then on, although not all of us realized it, regional councils would be more and more on their own, or dependent on their state governments, rather than the federal government, for assistance" (Scheiber, 1992, p. 21). As indicated earlier, the MWCOG gradually has come to rely more on contracts and less on federal support. Even after 35 years of the MWCOG's existence, its members seem to be unwilling to grant the body a more permanent commitment through larger dues. Instead, the MWCOG has been forced to become more entrepreneurial and more explicitly service-oriented in its relation to the local jurisdictions upon which it ultimately and increasingly depends.

In these respects, the story of DC's MWCOG pretty much mirrors that of COGs throughout the country. Atkins and Wilson-Gentry (1992, p. 475) point out that by 1992 only 13 of the 48 1970s-vintage federal programs promoting substate regionalism still were being funded. These authors indicate that rather than seeking to redirect the goals and decisions of their members, regional councils have adopted a guiding rule of emphasizing the distribution of rewards.

Figure 5.7 further illustrates the two key points we wish to emphasize: that the MWCOG's pattern of growth and decline reflects broad national policies and that it is, in this respect, more like than different from other regional councils across the country. The upper panel of the figure juxtaposes the proportion of the MWCOG's budget coming from federal grants with the percentage of all state and local expenditures coming from federal grants-in-aid during the same time period. The latter serves as an indicator of the nature of U.S. fiscal federalism and shows clearly the effects of Reagan's New Federalism policies, intended to decrease the role of the national government during the 1980s (see Conlan, 1988). The close fit between the two (Pearson's $r = .87$) suggests that the MWCOG's experience in large measure was shaped by broad political and philosophical trends. The lower panel, which juxtaposes the same measure of fiscal federalism with a simple count of the number of regional councils throughout the country, shows just as strong a relationship ($r = .83$).

■ Conclusions

The Washington metropolitan region is extremely inelastic, in Rusk's terms, and the prospects for changing this are not good. The Metropol-

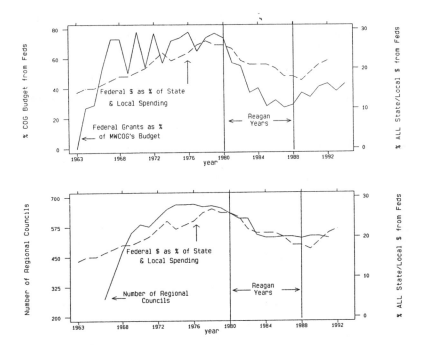

Figure 5.7. Regionalism as a Reflection of Fiscal Federalism

itan Washington Council of Governments (MWCOG) provides a mechanism for communication, information collection, and coordination among local governments, but its potential as a medium for metropolitan governance—or even as a forum for public debate and deliberation about important regional issues—is extremely limited. Its status as a voluntary, financially dependent organization forces it to shy away from controversy. Federal support, although undoubtedly contributing to the growth and institutionalization of the MWCOG, only serves to make these limitations more apparent. While Congress and the White House were aggressively seeking to bolster regionalism, the MWCOG was able to increase its visibility, build a professional staff and reputation for expertise, and earn the respect and, in some cases, gratitude of local elected officials. Even then, however, the MWCOG's ability to shape its own agenda was constrained. It was constrained by federal actors, who gradually tightened the strings attached to the decreasing financial support they made available. The MWCOG could not survive without federal support, but that federal support means that the organization's agenda is not set by local actors responding to local values, problems, and democratically expressed interests.

Formal dependency might not be so constraining if the MWCOG could depend on a strong constituency to rally around it when local politicians place parochial concerns above those of the region or when federal officials threaten to pull too tightly on the strings attached to their support, but the MWCOG's constituency is shallow. The benefits it provides are clearer to administrators and bureaucrats within the public sector than they are to elected leaders, and clearer to elected leaders than to the citizenry at large. Politicians can win votes by railing against MWCOG interference in local matters; they are unlikely to be held to account at the polls for failing to sacrifice local prerogatives in the name of an abstract metropolitan interest or loyalty to the MWCOG as an overarching institution. MWCOG's defenders can be proud of the organization's ability to survive the lean Reagan years, and they can make a case, too, that the organization has become somewhat leaner, more innovative, and entrepreneurial as a result. With greater reliance on contracting out, on member services, and on local sources of revenue, however, the MWCOG's leverage for shaping local governance seems to be shrinking.

Are the MWCOG's limitations inevitable, or do they represent a remediable failure of vision and political will? If it is true that local jurisdictions have a shared stake in overcoming regional inelasticity and fragmentation, as Rusk and others argue, it is possible that citizens can be mobilized around that common interest to become a reliable constituency for stronger regional institutions. Our consideration of the history and practice of regionalism in the Washington area reminds us of Mancur Olson's (1971) observations about the limited allure of shared interest as a spur to collective action, unless it is underwritten by selective benefits. The member jurisdictions are not drawn to the MWCOG by an overwhelming recognition that they are all rowing in rough waters in a single lifeboat. More significant seems to be their calculation that it is only through membership that they can share in certain federal grants and in the efficiencies gained through cooperative purchasing agreements and the like. Without ignoring the case for common metropolitan interests, it may be that proponents of regional solutions need to sprinkle their strategies with a realistic dose of sensitivity to the continued strength and immediacy of more parochial pressures and ties.

Given its organizational and political limitations, the MWCOG's role in the DC area is not likely to change dramatically in the near future. Federal programs such as ISTEA will still require monitoring and distribution of funds. The MWCOG's activities in providing interlocal services will likely continue to grow. What seems clear, however, is that the MWCOG will not lead the charge for regional government in the DC area. In 1993, the MWCOG formed the Partnership for Regional Excellence (PRE), a group of 200 business, government, education, labor, and community leaders organized to discuss the future of the metropolitan area and the MWCOG's long-term role in that future. One of the PRE's first actions was expressly to reject any notion of trying to establish a regional government. The PRE report called the very thought of regional government "extremist."

NOTES

1. Any laws passed by the DC city council can be overturned by the House District Committee, and Congress retains the power to conduct a line-item review of the city's budget. Congress's interest in exercising its power to intervene waxes and wanes, with concern about crime and the city's financial situation often being the precipitant of closer scrutiny and intervention. In 1994, with Congress paying close attention to the District's budgetary problems, there was a renewed interventionist spirit in Congress, with some even calling for a complete federal takeover.

2. Only Detroit, New York City, and Salem, Oregon, had COGs in 1957. Only three other cities—Los Angeles, San Francisco, and Atlanta—had COGs prior to the onset of the Great Society.

3. In 1982, the Department of Housing and Urban Development (HUD) dropped its Areawide Housing Opportunities Plan, which had provided bonuses as an incentive to encourage local jurisdictions to cooperate in the development of regional housing plans. Under the 1990 National Affordable Housing Act, federal housing grants are distributed directly to state and local jurisdictions rather than through regional organizations such as the MWCOG.

4. Such activities include assisting its members in applying for federal funds and preparing reports on the housing needs, resources, and strategies of its members. We emphasize that the MWCOG's shortcomings here are more reflective of structural constraints that limit all similarly constituted regional organizations than of lack of will or capacity of its leadership. Indeed, the MWCOG takes pride in the fact that it appears to be more aggressive than most other metropolitan bodies in tackling the controversial issue of low-income housing allocation. Its September, 1992, report on "Fair Share Housing Principles for the Washington Metropolitan Area" indicates that "it should be noted that in discussions with representatives from 10 of the major metropolitan area's [*sic*] regional councils across the country, not one other council with a federal fair share plan was identified" (p. 3). Moreover, the suburban jurisdictions are not unrelievedly hostile toward affordable housing.

Prince George's County allowed substantial numbers of units of subsidized housing during the 1960s, and since then both Montgomery County and Fairfax County have dramatically increased their shares of the region's affordable housing. These latter two jurisdictions seem to recognize that they have an interest of their own in providing some housing alternatives to middle-income and working-class families, who are an important part of their public and private employee base.

5. It should be emphasized that this can at best provide only a partial picture of participation in the MWCOG. Because the Police Chiefs' Committee is a relatively small part of the MWCOG, at least in terms of financial support, participation in the activities of this one committee should not be considered to be a reliable indicator of cooperation and participation overall. Even within the Police Chiefs' Committee, there are avenues for participation that our measure does not tap. Among these are attendance at training seminars, the level of participation in MWCOG programs, and attendance at subcommittee meetings, where the bulk of public safety-related business is conducted by the MWCOG.

6. The most frequent participants among the federal agencies were the Federal Protective Service (69), the U.S. Park Police (65), and the Secret Service (28 from the Uniformed Division, 18 from the Washington Field Office). Because the column representing federal attendance in Figure 5.4 is based on the aggregated attendance of many distinct federal agencies, it should not be compared to the columns representing other jurisdictions in order to make inferences about relative attendance rates.

REFERENCES

Atkins, P. S., & Wilson-Gentry, L. (1992). An etiquette for the 1990s regional council. *National Civic Review,* Fall/Winter, 466-487.

Conlan, T. (1988). *New Federalism: Intergovernmental reform from Nixon to Reagan.* Washington, DC: Brookings Institution.

Department of Metropolitan Development and Information Resources. (1991). *Population change in metropolitan Washington: A comparison of 1980 and 1990 census population and housing unit counts by subarea.* Washington, DC: Metropolitan Washington Council of Governments.

Derthick, M. (1972). *New towns in-town.* Washington, DC: Brookings Institution.

Gale, D. (1987). *Washington, D.C.: Inner-city revitalization and minority suburbanization.* Philadelphia: Temple University Press.

Grier, G. (1993). *The growth continues: An analysis of metropolitan Washington's 1990-1991 population change.* Washington, DC: Greater Washington Research Center.

Grodzins, M. (1966). *The American system.* Chicago: Rand McNally.

Harrison, R. J., & Weinberg, D. H. (1992, May). *Racial and ethnic residential segregation in 1990.* Unpublished revision of paper prepared for the meeting of the Population Association of America, Denver, CO.

Henderson, N. (1994, February 9). Kelly budget calls for tax increases. *The Washington Post,* p. A10.

Henig, J. R. (1985). *Public policy and federalism.* New York: St. Martin's.

Henig, J. R., & Gale, D. E. (1987). The political incorporation of newcomers to racially changing neighborhoods. *Urban Affairs Quarterly, 22,* 399-419.

Jaffe, H. S., & Sherwood, T. (1994). *Dream city: Race, power, and the decline of Washington, D.C.* New York: Simon & Schuster.

Ledebur, L. C., & Barnes, W. R. (1992). *Metropolitan disparities and economic growth.* Washington, DC: National League of Cities.

Metropolitan Washington Council of Governments. (1991). *Population change in metropolitan Washington: A comparison of 1980 and 1990 census population, by age, race, sex, and ethnic group.* Washington, DC: Author.

Office of Policy and Evaluation. (1992). *Indices: A statistical index to the District of Columbia.* Washington, DC: The District of Columbia Government.

Olson, M. (1971). *The logic of collective action.* Cambridge, MA: Harvard University Press.

Pianin, E., & Milloy, C. (1985, October 6). Does the white return to D.C. mean "The Plan" is coming true? *The Washington Post,* p. D1.

Rusk, D. (1993). *Cities without suburbs.* Washington, DC: The Woodrow Wilson Center Press.

Savitch, H. V., Collins, D., Sanders, D., & Markham, J. P. (1993). Ties that bind: Central cities, suburbs, and the new metropolitan region. *Economic Development Quarterly, 7*(4), 341-357.

Savitch, H. V., Sanders, D., & Collins, D. (1992). The regional city and public partnerships. In R. Berkman, J. F. Brown, B. Goldberg, & T. Milanovich (Eds.), *In the national interest: The 1990 urban summit.* New York: The Twentieth Century Fund Press.

Scheiber, W. A. (1992). *The Metropolitan Washington Council of Governments 1966-91.* Washington, DC: Metropolitan Washington Council of Governments.

Smith, S. (1974). *Captive capital: Colonial life in modern Washington.* Bloomington: Indiana University Press.

U.S. Bureau of the Census. (1991). *State and metropolitan area databook, 1991.* Washington, DC: Government Printing Office.

6

Louisville: Compacts and Antagonistic Cooperation

H. V. SAVITCH
RONALD K. VOGEL

In this chapter, we examine Louisville and Jefferson County, Kentucky, to understand how a metropolitan system of governance may evolve and to consider whether a metropolitan system of governance can substitute effectively for metropolitan government. Louisville and Jefferson County pursued an incremental and pragmatic approach to metropolitan reform that provides a test for the metropolitan governance thesis.

Louisville and Jefferson County have taken an intermediate approach to metropolitan reform. The two governments adopted a joint city-county compact in 1986, after failed efforts at city-county consolidation in 1982 and 1983. The compact directly addresses tax sharing, annexation, and joint agency management and funding. As a result of the compact, a new joint city-county economic development agency was established. The compact falls short of the one- and two-tier comprehensive government restructuring promoted by advocates of metropolitan government. The compact, however, goes far beyond the formal and informal interlocal government understandings and agreements envisioned by the public choice school. The compact exemplifies the metropolitan governance without government approach.

The compact led to the establishment of a strong public-private partnership in the community. This partnership has become institutionalized in a formal organization that ties the mayor, the county judge, and top business leaders of the community in a corporate-centered regime (Vogel, 1990, 1994). Thus, metropolitan governance in the Louisville region resulted from a restructuring of organizational relations among and between the central city and county governments and business (Kleinberg, 1995).

Although the compact reflects a shift in strategy from metropolitan government to metropolitan governance, there have recently been renewed calls for city-county consolidation or other reforms aimed at establishing formal metropolitan government. Renewed pressure for metropolitan government in Louisville and Jefferson County emanates from concerns over the community's ability to compete in the new global economy. A chamber of commerce strategic planning process identified reforming local governance as a major priority in economic development, leading to the appointment of a local commission to study and propose reform leading to metropolitan government for the community.

■ Defining the Functional City: A Regional Profile

The Louisville city-region roughly corresponds to the metropolitan statistical area (MSA). The boundaries of the Louisville MSA have changed over the years, reflecting changing levels of social and economic integration in the region. For purposes of this chapter, the Louisville region will be treated as Jefferson, Oldham, Bullitt, and Shelby Counties in Kentucky and Clark, Floyd, Harrison, and Scott Counties in southern Indiana (Figure 6.1).[1]

Population

The region has a population just under 1 million persons (see Table 6.1). The City of Louisville accounts for about 27% of the region's population. The central county—Jefferson County, including Louisville —makes up about 68% of the region's population. There has been a sharp decline in the city's share of the county population in the last four decades (see Table 6.2). Louisville's population dropped from 75% of the county in 1950 to 40% in 1992. The city's population has now stabilized, ending three decades of continued population loss. The 94 small cities hold 20.5% of the county's population, with the remaining 40% living in the unincorporated areas.

The Louisville MSA population will surpass 1 million persons by the year 2000. It will grow to 1,153,008 persons, a 17% increase, between

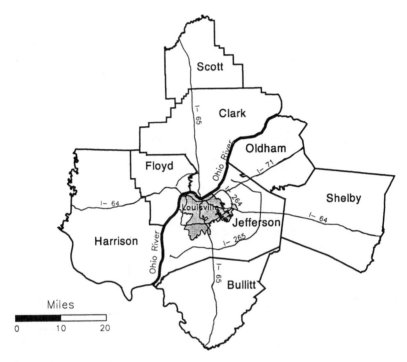

Figure 6.1.   Map 1: Map of the Louisville Region

1995 and 2020, pushed by a growing economy (Cornerstone 2020, 1994, p. 11). That population growth will increase the absolute population of Jefferson County but will reduce Jefferson County's share of the population from about 68% of the region to 63.5% of the region by the year 2020 (Cornerstone 2020, 1994, p. 20).

Economy

The regional economy is based in Louisville and Jefferson County. Figure 6.2 illustrates the transformation from a manufacturing-based economy to a service-based economy in Jefferson County. In 1969, 28% of all jobs in Jefferson County were in manufacturing. Projections are that only 8% of Jefferson County's jobs will remain in manufacturing by the year 2020. Service jobs, on the other hand, will rise from about 23% of

TABLE 6.1 Population of Louisville Region by County, 1950-1992

	1950	1960	1970	1980	1990	1992[a]
Jefferson	484,615	610,639	695,055	685,004	664,937	670,837
		+26.0%	+13.8%	−1.4%	−2.9%	+0.9%
Shelby	17,912	18,493	18,999	23,328	24,824	25,829
		+3.2%	+2.7%	+22.7%	+6.4%	+4.0%
Oldham	11,018	13,388	14,687	27,795	33,263	36,461
		+21.5%	+9.7%	+89.2%	+19.6%	+9.6%
Bullitt	11,349	15,726	26,090	43,346	47,567	51,128
		+38.5%	+65.9%	+66.1%	+9.7%	+7.5%
Harrison	17,858	19,207	20,423	27,276	29,890	30,737
		+7.6%	+6.3%	+33.5%	+9.6%	+2.9%
Floyd	43,955	51,397	55,622	61,169	64,404	67,194
		+16.9%	+8.2%	+9.8%	+5.3%	+4.3%
Clark	48,300	62,795	75,876	88,838	87,777	89,658
		+30.0%	+20.8%	+17.0%	−1.2%	+2.1%
Scott	11,519	14,643	17,144	20,422	20,991	21,572
		+27.1%	+17.0%	+19.1%	+2.8%	+2.8%
MSA Total[b]	635,007	791,645	906,752	956,756	952,662	971,844
		+24.7%	+14.5%	+5.5%	0%	+2%
Regional Total[c]	646,526	806,288	923,896	977,178	973,653	993,416
		+24.7%	+14.6%	+5.8%	0%	+2%

SOURCE: U.S. Census.

a. 1992 Sub-county Population Estimates, U.S. Bureau of the Census.

b. The MSA definition used is the 1990 designation, which included Jefferson, Oldham, Shelby, and Bullitt Counties in Kentucky and Clark, Floyd, and Harrison Counties in Indiana.

c. The regional total consists of Jefferson, Oldham, and Shelby Counties in Kentucky and Clark, Floyd, Harrison, and Scott Counties in Indiana.

county employment in 1980 to an estimated 41% by 2020 (Cornerstone 2020, 1994, p. 21).

Although experiencing many of the problems associated with deindustrialization, the Louisville economy has done well compared to other metropolitan areas around the country. The Economic Performance Index for Cities (EPIC), which weights job growth by wages, ranked the Louisville metropolitan area 15th out of the largest 75 metropolitan areas in 1993. More than 92,000 net jobs were created between 1980 and 1990, and 120,000 net new jobs are projected between 1995 and 2020 (Cornerstone 2020, 1994, pp. 5, 7).

TABLE 6.2   Population of Louisville as Share of County, 1950-1992

	City of Louisville	Percentage Growth	Jefferson County	Percentage Growth	City Population as Percentage of County
1950	369,129		484,615		76.1
1960	390,639	+5.9	610,947	+26.0	63.9
1970	361,706	−7.4	695,055	+13.7	52.0
1980	298,451	−17.4	685,004	−1.4	43.5
1990	269,063	−9.8	664,937	−2.9	40.4
1992[a]	271,038	+0.7	670,837	+0.9	40.4

SOURCE: U.S. Census.
a. 1992 Sub-county Population Estimates, U.S. Bureau of the Census.

The transformation to a postindustrial economy raises several concerns in the community. Many of the newer service jobs pay lower wages than the manufacturing jobs they replaced. In addition, job growth figures also include multiple job holders, as many workers now hold several part-time jobs. The Cornerstone 2020 (1994) report estimates that one third of the growth in jobs in the economy between 1980 and 1990 was due to multiple job holding (Cornerstone 2020, 1994, p. 8).

■ Regional Development

The Post-City Metropolis

The American pattern of metropolitan growth and development is noted for the presence of "economic inequalities" within and between the central city and the suburbs and the "auto-centered form of urban growth" (Angotti, 1993, p. 38). Growth and development of the Louisville region parallels that of other metropolitan areas in the United States. Decentralization of population and employment are apparent over the last several decades.

The uneven development associated with postindustrial cities is also present in the Louisville region. Population, housing, and job growth and decline are not evenly distributed throughout the community. For example, while the city population was declining by 27% between 1970 and 1990, the population of the county outside of Louisville was growing by 18% (see Table 6.2).

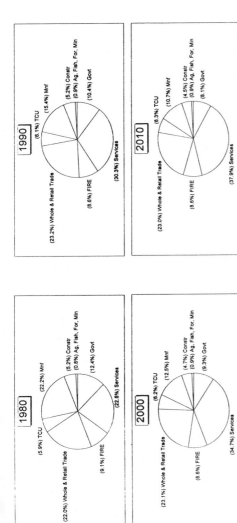

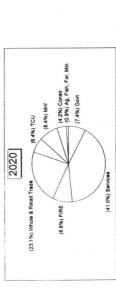

Figure 6.2. Jobs by Major Industry, Jefferson County, 1980–2020

SOURCE: Greater Louisville forecasts of jobs, population, and income: 1995-2020. *Cornerstone 2020*, Final Report, July 1994, p. 21.

TABLE 6.3 African American Population, Louisville and Jefferson County

	Louisville			Jefferson County		
	African American	Total	Percentage	African American	Total	Percentage
1950	57,772	369,129	15.6	62,750	484,615	12.9
1960	70,449	390,639	18.0	78,890	610,947	12.9
1970	86,040	361,472	23.8	95,588	695,055	13.7
1980	84,080	298,451	28.1	109,702	685,004	16.0
1990	79,783	269,063	29.6	113,435	664,937	17.0

SOURCE: U.S. Census.
NOTE: For 1950 and 1960, numbers are for "nonwhites." For 1970, 1980, and 1990, numbers are for African Americans only.

African Americans are still concentrated in the City of Louisville (see Table 6.3). There is some evidence of black suburbanization on a small scale. African Americans living in Jefferson County outside the City of Louisville have risen from only 3% of the population in 1970 to 9% in 1990. Most of the suburbanization, however, represents expansion of existing African American neighborhoods in the West End, a section east of downtown, and the Newburg neighborhood outside the city—which has a 52% African American population. In other words, black suburbanization is not leading to a change in the segregated housing patterns (Adams & Bartlett, 1995).

By 2020, Central Louisville, the portion of the city that includes the downtown, is projected to lose 10% of its population, stay about even in housing, and gain about 9% in jobs. The West End in the City of Louisville, home to most of the region's African American population, is expected to experience a 21% drop in population, 13% decline in housing units, and 28% loss of jobs (Cornerstone 2020, 1995, pp. 4, 5, 6). Population, jobs, and housing are continuing to shift to the newer, more affluent suburbs on the eastern side of the county (Cornerstone 2020, 1995, p. 1).

The central city and county remain the dominant force in the regional economy and show little sign of imminent or long-term obsolescence. Jefferson County will retain a significant share of the MSA employment in the future. The county is projected to hold 75% of the jobs in the MSA in the year 2020, down only 6% from its 81% share in 1970 (Cornerstone 2020, 1994, p. 21). Today, 93% of Jefferson County residents work in the county, and residents make up 80% of all workers in the county (Price, Coomes, & Hoyt, 1995). The downtown area continues to serve as the legal-political and finance, corporate-administrative center of the region. The downtown area is expected to gain an additional 20,000 jobs by 2020 (Raymond, 1992).[2]

Thus, the Louisville region is representative of the maturation of the American metropolis in the post-city era. Although experiencing the pains and difficulties associated with deindustrialization and decentralization, the core has maintained its central functions. It is the site of a highly developed physical infrastructure and urban service delivery system, providing a locale for economic growth and investment in the region, cultural and intellectual stimulation, and a diverse set of neighborhoods for residents to experience changing lifestyles. At the same time, the core will never be as preeminent in this new post-city era, with a population dispersed throughout the region and the development of edge cities and nodes of activity connected by an interstate highway system and beltway that ties suburbs, small cities in the central county, and outlying counties into the metropolitan system. The inequalities that characterize the new American metropolis, however, also exist in the Louisville region.

Regional Development Strategies

The community has attempted to restructure its economic development efforts over the last decade to make them more effective. This involved organizational restructuring, community goal setting, and a Regional Economic Development Strategy that included an emphasis on infrastructure projects and local government reorganization.

In 1988, a public-private partnership for economic development was forged and institutionalized in the Partnership for Greater Louisville. The mayor and the county judge serve as vice-chairmen of the Partnership, and the two governments provided $500,000 for the annual operating budget of the Partnership as well as contributing $1 million toward the initial $10 million capital fund drive for business attraction incentives. The Partnership was to take the lead role in economic development and coordinate activities with the Chamber of Commerce and the Office of Economic Development (OED) (Ehrenhalt, 1989; Vogel, 1990).

The actual economic development strategy of the community was less clearly defined. An earlier effort at strategic planning, Forecast Louisville, undertaken in 1987, had been repudiated by its own sponsor, the Chamber of Commerce, without explanation. Now that the organizational structure of economic development had been settled, it was time to decide on a strategy of economic development. At the time, the strategy appeared to be focused on infrastructure development, most notably a $300 million airport expansion to benefit the United Parcel Service (UPS)[3] and an ad campaign oriented toward community image building.

The original aspirations for the Partnership were not achieved. The Partnership failed to develop an economic development plan for the community and focused its energies on business attraction, marketing the community to selected industries identified by a consultant. Its role as coordinator and planner of the community's economic development effort was never realized. Economic development efforts continued to be ad hoc, driven by individual agendas and emphasizing business attraction efforts and infrastructure projects such as building a bridge, constructing a stadium, and developing the waterfront (Negrey & Vogel, 1991).

By 1990, a committee recommended creating one economic development agency that would consolidate the Chamber of Commerce, the Partnership, and the Office of Economic Development (OED) as well as chambers of commerce in the surrounding counties (reported in the *Courier Journal,* August 23, 1990, p. 1B).[4] The director of the OED, upon resigning his position, gave his assessment of economic development to a local leadership group, complaining that there was a lack of coordination among the community's economic development organizations and that the community had no vision or direction of where it was trying to go. He also pointed to problems with arranging project financing and an under-qualified workforce (reported in the *Courier Journal,* October 17, 1990, p. 1B).

The proposal to create a single economic development entity did not have much support. This reconsideration of the organization of the community's economic development efforts and the vacancies in the Office of Economic Development, Chamber of Commerce, and Partnership did mean that a new team was put in place and greater thought was given to the direction in which economic development needed to go. This led to two simultaneous agenda-setting processes in the community. One was a broad-based goal-setting effort that was to build a regional consensus on community issues. The other was a more focused economic development strategic planning process that would be folded into the goals project as one of the elements.

The goal-setting process, called Goals for Greater Louisville, although certainly broad-based, was really more of a public relations venture. It was modeled after the Goals for Dallas and partly intended to help heal the community following a controversial decision to relocate 3,500 residents for airport expansion (see Note 3). A number of committees were formed, and citizen leaders worked hard to study the issues and forge consensus. A set of goals was submitted to the citizenry by printed ballots published in the local papers or distributed to community groups for voting. Community groups tried to improve the prospects of certain projects by encouraging members to fill in ballots.

Important issues under consideration in the community, such as busing for desegregation, were mostly excluded from consideration by the group. It is interesting that one of the most controversial issues revolved around building a bridge across the Ohio River to southern Indiana. The transportation committee did not support the building of a bridge, an important economic development priority in southern Indiana. Nor did the Regional Economic Development Strategy (the economic element of the goals project managed by the Chamber of Commerce) include the bridge as a priority. The bridge issue, however, did appear on the ballot, put there by the regional cooperation committee.

The final set of adopted goals was disparate, inconsequential, and quickly forgotten. Leaders never bought into the process except as a device to demonstrate concern for citizen input and community building. The real action was on the economic development strategic plan.

The Regional Economic Development Strategy

When leaders concluded that it was not politically feasible to unify all economic development organizations into one agency, they were still left with two problems: how to bring about greater coordination among lead economic development organizations and the absence of a clear economic development strategy that everyone would buy into. The solution was to develop a set of policies or guidelines on which all could work together. The vehicle became the Regional Economic Development Strategy (REDS). It was decided that the Partnership would fund the strategic plan and that the Chamber of Commerce would house and manage the project.

The REDS sought to involve a more representative set of participants in economic development planning than in the past. Efforts were made to include geographic and demographic sectors of the community that were usually underrepresented, including African Americans, residents of the south side (a blue-collar area), and women.[5]

Over a 2-year period, a strategy for economic development was forged, culminating in six objectives for implementation published in April of 1993 in a document titled *Investment Strategies* (Regional Economic

Development Strategy, 1993). The overall mission that guided the process was to

> [g]enerate sustained economic growth of the Metropolitan Statistical Areas (MSA) exceeding the national average (identified by the number of jobs created and per capita income) and equitably distribute the benefits of this economic growth to all demographic groups and geographic areas of the region. (Regional Economic Development Strategy, 1992, p. 1)

The final six priority objectives selected for implementation to achieve this goal were chosen based on the willingness of Partnership donors to fund these objectives. The objectives that were selected were to

1. study and reorganize local government in Jefferson County,
2. improve the quality of the workforce,
3. strengthen minority economic development,
4. initiate and provide financial support for a comprehensive land use and infrastructure plan,
5. increase the flow of federal and state aid to the Louisville metro area, and
6. improve coordination of economic development activities and organizations in the community.

Unlike many such projects in other communities, implementation has not been left to chance. The REDS staff and executive committee, including representatives of the mayor and county judge-executive, continued to meet and work on implementation. The Partnership funds the effort with a contract to the Chamber of Commerce, which manages the project. This ensures that REDS is a major component of their agency work plans. The Partnership has agreed to raise money to carry out these goals, although some goals are higher on the agenda than others.

With respect to regional development, the most important goal is the writing and adoption of a comprehensive plan, known locally as Cornerstone 2020.[6] Many of the development issues that might normally have been expected to be part of a regional economic development strategy actually will be dealt with in the comprehensive plan itself.

The Partnership, through the Chamber of Commerce, is financing upwards of half of the $3,520,000 that has been budgeted for the comprehensive plan. The Chamber of Commerce is directly responsible for managing the marketplace component and community form component of the plan as well as developing the regional perspective that will inform the plan. This is a sign of the strength of the public-private partnership in the community and the degree to which the two sectors have not only coordinated their efforts but also permeated each other.

■ Regional Governance

Political Structure

The Louisville region, as defined, contains 286 local governments, including 8 counties and 132 cities. Focusing only on Jefferson County, which holds almost 70% of the region's population, there are 126 local governments (U.S. Census of Governments, 1992: Table 28). Of these, 96 are general-purpose governments, including one county government and 95 city governments.[7] The other 30 are special-purpose governments in Jefferson County, including 2 school districts (1 countywide school district and 1 small independent city school district) and 28 special districts (these include 22 volunteer fire districts, a sewer district, and a transportation district).

Local government structure in Jefferson County actually is more fragmented than this, in that there are at least another 22 subordinate local government organizations that are not classified as separate local governments. These include joint city-county agencies under interlocal agreements, independent public agencies, authorities, and corporations. Whether this constitutes a fragmented local governmental system depends on one's perspective.

Interlocal Conflict and Cooperation

In the 1970s and 1980s, the city of Louisville and Jefferson County battled each other over annexation, joint agency operation and funding, and economic development. As in metropolitan areas throughout the country, population growth was occurring on the urban fringe, businesses were closing in the central city, and new business frequently was locating in the outlying areas. The city was feeling the pinch of less federal aid, a declining economy, and a loss of population, which resulted in less revenue at the same time that demands were being made by residents for more government services. City-county consolidation was proposed as a solution to these problems in 1982 and 1983. Both times, merger was rejected in public referendums. The city then pursued annexation with renewed vigor in the summer of 1985, proposing to annex all the unincorporated area in the county that was contiguous to the Louisville city limits.

Annexation seriously threatened county revenues. The county and the city both relied on the occupational tax for more than half of their revenues.[8] Employees inside the corporate boundaries of Louisville do not pay occupational taxes to Jefferson County. Therefore, every successful annexation by the City of Louisville reduces county revenues. For example, when the city annexed a suburban shopping mall, the county lost several hundred thousand dollars in annual occupational tax receipts.

Competition for economic development is also related to the occupational tax funding system upon which both governments rely. The city wanted new industry or business expansion to take place within city boundaries to gain the additional occupational tax revenues. The county, on the other hand, would not gain any additional occupational tax revenues if a business located or expanded in the city boundaries. Therefore, the county had an incentive to recruit business to sites outside the city limits.

By 1985, city-county disagreements over annexation, the financing of services in the joint-agencies, and more intense competition for new industry were leading toward a crisis. Legislators made it fairly clear that they would deal with the issues if the city and county did not, possibly by taking away the city's ability to annex unincorporated areas.

An opportunity to deal with these issues presented itself, in the fall of 1985, when outgoing city mayor Harvey Sloane was elected county judge-executive. For the first time, a mayor of Louisville would serve as county judge-executive. Sloane knew intimately the city's concerns. Sloane was replaced as mayor by Jerry Abramson, a former member of the Board of Aldermen. The two were both Democrats, shared the same concerns, drew on similar political support, and were on friendly terms. Over a period of several months, during the transition from election to taking office, the mayor and judge-executive were able to negotiate the Louisville-Jefferson County Compact.

The mayor and the county judge were striving to foster greater cooperation between the two governments, especially with respect to economic development and providing a respite for the community, which was torn by battles over merger, annexation, and intergovernmental warfare over economic development and joint agency operations. The compact was, in effect, a "peace treaty" between the two governments.

A number of factors contributed to the adoption of the compact.

1. Both top elected officials were of the same party, and the newly elected county judge-executive had just vacated the position of mayor of the city.
2. The state legislature had threatened to intervene to settle the problems if the city and county did not somehow resolve their differences.
3. The precedent of a number of joint agencies since the 1940s provided the governments with some experience at coproduction of services and cooperative interaction.
4. The business community was pushing for an end to competition for economic development that was perceived as harmful to Louisville's economic development efforts.
5. The failure of two previous merger efforts meant that leaders had to seek some alternative to city-county consolidation.
6. Battles over annexation were a no-win situation for both governments.
7. The mayor and judge-executive were able to get the compact adopted quickly, limiting the ability of groups to mobilize against the agreement, if they even were aware of it.[9]
8. An atmosphere of crisis pervaded the community, resulting from fierce competition for economic development and battles over annexation between the city and county.

The Louisville-Jefferson County Compact

If a cooperative relationship were to be established between the city and the county, the institutional and structural bases for the conflict and competition would have to be addressed. The root of the problem was the transformation of the economy, with accompanying deindustrialization and related decentralization of population and industry and declining federal aid. The fragmented local government structure fostered competition rather than cooperation between the city and county over jobs and revenues.

Conflict between the two governments was driven by their separate needs for adequate revenue sources, which led to an unhealthy competition for economic development and an unwillingness to cooperate to better provide services to the community. Further, differences in the priorities of the two governments and their financial abilities meant that the independent agencies that were funded by both the city and the county could easily play one unit against the other, exacerbating tensions. The city and county needed to change the rules so that cooperation rather than conflict was in their mutual interest.

The city and the county were able to resolve many of their differences with the adoption of the Louisville and Jefferson County Compact in 1986. The compact, set up to last for 12 years, expires in 1998. Because the agreement was adopted by the City of Louisville, Jefferson County, and the Commonwealth of Kentucky, no party can unilaterally withdraw, and the terms of the agreement cannot be altered until its expiration, at which time the compact can be renewed, be renegotiated, or simply expire.

Three major sources of tensions between the city and the county are addressed in the compact. First, the city and the county agreed to share revenue from the occupational tax. The City of Louisville and Jefferson County agreed to share the occupational tax based on a formula in which the city receives approximately 58% and the county 42% of the total revenue. Actual occupational tax collections by each government in 1985 served as the baseline to arrive at this formula and distribute revenues between the two governments. This forms the basis for ending divisive annexation battles and competition for economic development.

Second, in the compact, the governments agreed to new arrangements for funding and managing the independent joint agencies, designed to bring about greater political accountability and more fiscal control, but also involving a re-sorting of services between the two governments. Prior to the compact, the joint agencies were funded on an equal split between the city and the county and operated under an independent board, which appointed an executive director. Four agencies were reassigned to the county—air pollution, health, crime commission, and planning. Four were reassigned to the city—disaster and emergency services, human relations commission, history and science museum, and the zoo. Four remained joint agencies—library, parks, transit, and sewers. The city and county continue to split equally the costs of the remaining joint agencies. The executive director, however, is now appointed jointly by the mayor and county judge-executive, and agency boards became advisory, greatly reducing their independence. One new joint agency, the Office of Economic Development, was created under the terms of the compact.[10]

The third major part of the compact imposes a moratorium on annexation or new incorporations in Jefferson County. The city agreed to relinquish its efforts at annexation in return for the county assuming a greater share of the costs of providing urban services. In the re-sorting of services, the county agreed to take on an additional $1 million in service costs in return for the city relinquishing annexation efforts and to reflect the city's population decline, from 50% of the county population in 1970 to about 40% at the time the agreement was made. If the compact is not renewed or otherwise is terminated, the city retains the right to annex all lands it had attempted to annex prior to the compact's adoption.[11]

■ Assessing the Compact

The compact entails more than an agreement on how the city and county produce, deliver, and finance services in Louisville and Jefferson County. The compact represents a more comprehensive approach to inter-local agreements than the ad hoc, year-by-year incremental approach. There is not a single document that encompasses the entire agreement. A number of city and county ordinances were adopted to implement the compact. The compact must be understood not as a single document that binds the two governments together but as a process that institutionalizes cooperation between the two governments. An intangible aspect of the compact that is frequently pointed to is a spirit of cooperation that is said to bind the two governments. In the past, many community leaders treated the compact as sacred, although it is doubtful that many members of the community are even aware of the existence of the compact.

Benefits of the Compact

The compact has had several positive effects in the community. It appears to have led to some limited cooperation between the city and county governments that previously was lacking. Conflicts over annexation have been eliminated. The parceling out of eight previously joint agencies to one or the other government has reduced conflict over joint agency funding, and thus the overall level of tension between the city and county governments has subsided. Local officials also believe that the compact has been a boon to economic development, citing the extension of the city's enterprise zone to the county and the opportunity for a joint approach to economic development (Louisville/Jefferson County Office for Economic Development, 1990).

Since 1986, city and county government, under the guidance of the mayor and county judge, have forged a strong public-private partnership with local business leaders (centered in the Greater Louisville Economic Development Partnership) to set economic development priorities and coordinate public and private economic development efforts. One of the greatest successes of this partnership has been the airport expansion project to serve UPS. Without the compact providing the basis for a unified economic development effort by local government, it is unlikely that this project could have been accomplished.[12] The tax sharing provision of the compact has eliminated the intense competition between the two governments for economic development.

Criticism of the Compact

County budget problems in the early 1990s illuminated several weaknesses in the compact as a mechanism to keep the peace between the city and the county. Budget shortages caused the county to raise several objections to the compact. First, new growth in the community has been outside the city, leading the county to believe it is being shortchanged under the tax sharing formulas of the compact. Since the adoption of the compact 10 years ago, the county has transferred just under twenty million dollars to the city under the tax sharing agreement.

County officials also complain that the transfer of services between the city and the county was one-sided and that the city was able to shift the burden of more expensive services to the county without looking more broadly at service burdens not previously covered by joint agencies, such as county corrections. They argue that the county is left with services that are more costly and for which costs are increasing at a greater rate.[13]

Additionally, in the late 1980s, state legislation allowed Jefferson County to begin imposing a tax on insurance premiums sold within its boundaries. This tax had been in place in Louisville and other cities for several years. In Jefferson County, most small cities chose not to collect the insurance tax until 1990, when the county announced its intention to use the tax as a way to ease its revenue shortfall. A municipal tax on insurance premiums preempts the county tax. Small cities reasoned that if the tax was going to be collected anyway, they might as well keep the revenues in their own communities. County officials believe that hundreds of thousands of dollars are being siphoned off by small cities each year. They also believe that much of the revenue generated is inappropriately going to the City of Louisville because insurance companies have trouble determining to which government they should pay the tax. Jefferson County has called for an audit of these monies to ensure that it gets its fair share.

Underlying many of the county's complaints is the feeling that the compact really represents an agreement between two mayors—incoming

Mayor Jerry Abramson and outgoing Mayor Harvey Sloane, newly elected to the office of county judge-executive. County officials believe that Sloane did not adequately protect the county's interests.[14]

These issues came to a head in 1990, when Judge-Executive David Armstrong took office and found what he believed was a fiscal crisis requiring budget cuts in county expenditures. At one point, he proposed up to $1.5 million in joint-agency budget cuts alone, with another $1.4 million in proposed cuts to the health department, previously a joint agency that was now funded by the county. These proposed budget cuts posed a dilemma for the city. The compact provides for an equal funding to joint agencies. The city could cut back its contributions to joint agencies, to county funding levels; provide additional funding to joint agencies for additional services limited to city residents; or make up for county budget cuts with city funds to maintain the joint agencies' budgets and services. None of the options was particularly desirable, although there was some precedent for earmarking contributions. For example, in agencies such as parks and economic development, the governments have funded special projects of their own in the past.

Fear of a declining revenue base in the city and the county raised the specter of renewed city-county conflict and threatened the compact. Not only was joint-agency funding once again problematic, but 1990 census figures pointed to continuing population losses in the city (9.8%) and county (2.9%) between 1980 and 1990. At present, both the city and county population appear to have stabilized, and economic growth has meant that the revenues of both governments have grown.[15]

The "spirit of cooperation" between the city and county has been strained not only by arguments over money but also by disagreements over community priorities. The mayor and judge-executive have clashed over the question of building a new bridge connecting to southern Indiana. The judge favors a bridge in the eastern part of the county to complete an outer beltway and to promote economic development. The mayor favors a downtown site to relieve congestion and to rebuild a poorly designed intersection of three major interstate highways. Underlying their opposing positions are the divergent constituencies the two leaders represent, one based in the older central city and the other in the suburbanizing areas. This issue has regional implications as well, as southern Indiana economic and political leaders have indicated that an eastern bridge site is a pre-requisite to any future regional cooperation.

The mayor and the judge-executive have disagreed on other economic development priorities and state legislative funding priorities. For example, in 1994 the city's top priority was expansion of the downtown Commonwealth Convention Center, whereas the county wanted a new Hall of Justice. The University of Louisville had a separate agenda, and a private business group was pushing a stadium project as yet another priority. In an effort to compromise, all these projects were listed as top priorities in different categories. None was approved by the state. Given the legislature's surprising fiscal restraint, it is not clear whether any would have been funded even if governments and community leaders agreed on the community's top priority.[16]

The Future of the Compact?

This difference over city and county priorities, in conjunction with the county fiscal difficulties, has called the compact into question. Until recently, both city and county leaders spoke in glowing terms about the compact and refused to identify specific tensions between the governments. City and county conflicts did persist after the compact's adoption, and the relationship between the mayor and the judge is not as strong as it once was. There are questions about whether they consult directly with each other, although their staffs still work together closely. If no action is taken by the city, the county, and the state legislature to renew the compact prior to July 1, 1998, it is unclear what the consequences will be. In theory, things will return to the way they were in 1986, prior to the compact.

The mayor and county judge have indicated that they still favor some form of metropolitan government, preferably city-county consolidation. Both officials have stressed the need for one leader in the community who can set the agenda. This view has been echoed by the business community, which has argued that local government restructuring is critical for community economic development. The business community has been responsible for the major push to put merger back on the community agenda, growing out of a Chamber of Commerce strategic planning process, financed by the Greater Louisville Economic Development Partnership (Partnership) to develop an economic development plan for the community.

A local government task force was set up, but after several meetings, the group suddenly stopped meeting without explanation, even to several of the participants. The primary problems appeared to be that the Partnership had not yet made a commitment to finance and support the effort and local government reorganization was not ranked as highly as an economic development priority as other items by the Partnership board. This was rectified when, in April of 1993, the Partnership agreed on a final set of six priority objectives for implementation (Regional Economic Development Strategy, 1993). First on the list was to study and reorganize local government in Jefferson County. Still, business leaders were not in complete agreement that local government reorganization should take precedence over other economic development priorities, and further action required waiting until the Partnership raised funds to support this effort.

In January of 1994, there was movement on local government reorganization once again, this time emanating from the mayor and county judge. The mayor was clearly interested in city-county consolidation, not wanting to set up a process that might not result in this kind of proposal. The county judge was less clear, making reference to the possibility of other metropolitan government schemes. Fearing that the process was moving ahead without them, the City of Louisville Board of Aldermen and the Jefferson County Fiscal Court proposed their own study of metropolitan government.

After quiet and fierce negotiations among the mayor, the county judge, the president of the Board of Aldermen, and a Fiscal Court member,[17] agreement was reached on a new approach. A local government commission would be set up and given free rein to study local government, the existing compact, and alternatives to providing urban services, including possibly merger or metropolitan government. A 16-person nominating committee was set up, and the project coordinator for the Regional Economic Development Strategy was appointed to staff the committee pending the selection of the larger citizens' task force with 128 persons.

A formal and independent local government commission was not set up. Instead, the nominating committee also acted as the steering committee for the project. The Partnership continued to provide the financing of the project, which was still housed within the Chamber of Commerce. Early on, the committee was plagued by distrust among members representing various constituencies, including the city, the county, small cities, and the African American community. It was delayed several times in adopting a work plan, selecting the larger citizens' task force, and deciding what role the larger committee would play in the process.[18] A major mover behind the project resigned from the committee in October 1994, expressing concern that not enough progress was being made and that his presence was perhaps an obstacle, given his past preferences for merger (McDonough, 1994).

Since then, additional staff members were hired and the citizen task force was appointed. A series of educational sessions took place in the spring of 1995, and the group broke into smaller working committees to make recommendations. These groups released a set of draft recommendations in October 1995. A set of final recommendations was released in January 1996.

Merger was rejected:

> After careful consideration, it is the unanimous recommendation of the Study Group that City of Louisville should not be merged out of existence in order to accomplish delivery of county-wide services. Furthermore, the Study Group recommends to the Task Force that African American representation is a major issue for any governmental structure which may be selected by the Task Force. (Finance, Compact and Governmental Structure Group, 1995, p. 7)

Instead, the thrust of the Study Group's recommendation is that county government be modernized and that the community incrementally move toward regional urban service delivery, with countywide services including "police protection, emergency medical services, disaster emergency services, parks and recreation, public works, human relations, public housing, indigent care, the library system, air pollution, land use planning, economic development, TARC, MSD, and the Airport Authority" added to county responsibilities, in addition to corrections and public health, which the county now provides (p. 7). Shifting these services to the county also would require the accompanying occupational tax revenue (Jefferson County Governance Task Force, 1996). The city would be left an empty vessel, with minimal services to perform. In all likelihood, this proposal will be unacceptable to the city.

The final recommendations released in January 1996 also called for a county council form of government (county version of mayor council) with an enlarged 12-person county council elected in nonpartisan, at-large elections (nominated in districts). Task force leaders sought to have a bill introduced in the 1996 legislative session to hold a county referendum in the November 1996 general elections on changing the structure of county government. A version of the bill passed the Senate but was not acted on in the House.[19]

In the end, the local governance study project probably will leave the community where it was before the project began, that is, with the city and county renegotiating the compact's extension. There is no crisis suggesting greater reform or restructuring, and the only real consensus is the need to continue the compact in some form.

Some difficulty was experienced by the local government study group, resulting from the structure and process being followed and the history of local government reform in the community. Successful local government study and reorganization usually requires study commission members to be broadly representative of the community and to have some degree of independence to study issues and come to their own conclusions about the need for local government reorganization (National Academy of Public Administration, 1980).

The task force was fairly representative of diverse segments of the community and of pro- and anti-merger positions. However, the project was housed and managed by the Chamber of Commerce. There has always been some suspicion of how independent the process was and little evidence was offered in support of the final set of recommendations.

The Impetus for Local Government Reorganization

It is unclear whether the impetus for local government reorganization in the community really stems from economic development concerns or if concern for economic development has become the vehicle to sell local government reorganization.[20] The business community has long favored merger in this community. It is also unclear whether it is really economic development or some other issues that lead the mayor and county judge to favor merger or other metropolitan government schemes.

The most common explanation provided by both business leaders and the mayor and judge for local government reorganization is that a community must have one common vision and leader and that this is not the case in Louisville and Jefferson County. When they are asked for specific examples, few are forthcoming, except perhaps that the city and county have different sets of priorities, such as the location of another bridge across the Ohio River.

It is interesting that the bridge example is provided to illustrate the problem of having two "chief executives" in a community. Even with a single executive, consensus would still be elusive on the bridge issue. The Chamber of Commerce itself has been unable to make a recommendation on the bridge because of differences among its members on this issue. The issue is conspicuous by its absence in the Regional Economic Development Strategy.

The call for a single vision in the community rests on the assumption that there is something inherently wrong with the city and county having potentially conflicting goals. In fact, it is natural for them to have conflicting goals, as the governments represent different political constituencies. The city government represents an inner-city population including about one third African Americans. The city has older neighborhoods, a higher level of poverty, and greater diversity in its population, and it houses the central business district. This leads to differences with respect to needed services and a greater concern with the revitalization of the downtown area.

Although the county has the same concerns, to some extent, it also is concerned with the needs of the 60% of the county population not in the City of Louisville. It is more focused on infrastructure placement in the outlying areas because these are the areas for which it is primarily responsible. Although differences between the two governments may sometimes result in differing priorities in terms of state legislative funding, it is usually possible to compromise on issues. The existence of the compact, particularly its tax sharing provisions, greatly facilitates cooperation among the two governments.

The notion exists that somehow, if the city and county governments were to consolidate, there would be only one community vision in Jefferson County. There still would be central city needs separate from those of the rest of the county, given the population, housing stock, age of the infrastructure, and sunken investment in the downtown area. The City of Louisville only holds 40% of the county population, but central Louisville (i.e., the downtown area and immediately surrounding neighborhoods) holds only 5% of the population. A single vision may be enunciated from the newly consolidated government, but in all likelihood it would be a suburban vision, without the existing central city political institutions to focus attention and resources on unique central city concerns.

■ The Compact: Metropolitan Government or Governance?

Is the compact a viable alternative to metropolitan government, providing for metropolitan governance without formal metropolitan government? Does areawide coordination and integration occur in the delivery of urban services? Is there a need to establish an umbrella body that can view matters from a metropolitan perspective and act on behalf of a metropolitan interest?[21] In some respects, the answers to these questions depend on whether one defines the metropolitan region as Jefferson County or as the Louisville Metropolitan Statistical Area plus Shelby County.

If the region is defined as the Metropolitan Statistical Area plus Shelby County, then coordination and integration are practically nonexistent. The only multicounty entity is the Kentuckiana Planning and Development Agency (KIPDA), which serves as the metropolitan planning organization for transportation issues. The KIPDA, however, does not act as a council of government and does not serve as an umbrella body to address metropolitan issues.

If Jefferson County is treated as the metropolitan area (holding about 70% of the region's population), then the conditions for metropolitan governance exist in some measure. Although there is not a single metropolitan government in the county, the compact essentially brings together several functions—economic development, transit, libraries, parks, and sewers—under a single countywide authority. Although there is not a unified government, the mayor and county judge acting in tandem oversee administration of these agencies and appoint their executive directors.[22]

Further coordination occurs between the city and county through regular contact between their top staff members and participation by the two governments in a strong public-private partnership with the business community. Underlying the coordination between the two governments is an agreement about sharing tax revenues from the occupational tax and a moratorium on city annexation that provides the basis for cooperation. Although not an ideal arrangement from the standpoint of advocates of metropolitan government, this is far better than that found in many communities.

The public-private partnership includes the top political executives (mayor of Louisville and judge-executive of Jefferson County) in the central county and the top business leaders (major employers, banks, etc.). A bureaucratic structure has been created in the form of the Partnership for Greater Louisville to manage and nurture this relationship. The Partnership finances economic development incentives and underwrites community projects that tie into urban economic development. Many of its decisions are implemented by way of contracts with public and private agencies, including the Chamber of Commerce and the Louisville/Jefferson County Office of Economic Development.

The Partnership for Greater Louisville constitutes the core of a corporate centered regime (Vogel, 1990). There is no doubt that the leaders of this regime direct the community's economic and urban development strategy. The regime, however, does not extend beyond the boundaries of Jefferson County. The effort to extend the regime to include southern Indiana was rebuffed when the top development issue of southern Indiana, a bridge completing a beltway, was not included in the Regional Economic Development Strategy (Copas, 1995).

■ Conclusion

Urban politics now extends beyond the central city into the suburbs and "exurbs," which now make up the metropolis. Regional politics is an extension of urban politics. Leaders in the Louisville region are attempting to adapt, evolve, or invent institutions, processes, and policy to better respond and shape the social, economic, and political forces that act on the community. The definition of the community and region, however, are themselves in contention.

In the Louisville region, there is evidence that the central city and county have evolved an "antagonistic cooperation" to bring some modicum of metropolitan governance without government. Citizens and leaders may have accepted some enlargement of the region beyond the central city, but many subsets of the community (e.g., suburban cities, minorities, and central city residents) are resistant because of real and imagined effects on their group or individual interests. It has proved even more difficult to develop a larger regional identity that includes the remaining counties of the MSA. Although economically dependent on the core, these counties have been reluctant to place themselves in a politically subordinate position in relation to the center. Multicounty regional institutions are weak or nonexistent.

In the case of the Louisville region, this lack of a larger regional identity is not as serious as it otherwise might be, in that well into the 21st century, the central county will dominate the region economically. It potentially could be more serious in that it may threaten investment in costly physical infrastructure needed in the region (e.g., a bridge across the Ohio River), and high commuting levels and air pollution mitigation efforts could eventually place severe limits on economic development opportunities in the region.

NOTES

1. This is not the current U.S. Census metropolitan area designation. In 1950, the Louisville metropolitan area consisted of Jefferson County in Kentucky and Clark and Floyd Counties in Southern Indiana. In 1970, Oldham and Bullitt Counties in Kentucky were included in the MSA. For the 1990 census, Shelby County in Kentucky and Harrison County in Indiana were added to the MSA.

Following the 1990 census, the U.S. census dropped Shelby County from the MSA and added Scott County, Indiana. Today, the census-designated MSA no longer conforms to leaders' and citizens' notions of what constitutes the region. For all practical purposes, Shelby County is treated as part of the Louisville region. The Bureau of Economics at the University of Louisville, which monitors the local economy, continues to include Shelby County in its reports on the Louisville regional economy. Few citizens realize that Scott County, Indiana, was added to the MSA.

2. There was some concern in the 1980s that the core would lose its centrality. In 1980, "there were 8,000 more people commuting out of the Louisville area than were commuting in." By 1990, this had reversed, with 13,000 more persons commuting into Louisville than out (Cornerstone 2020, 1994, p. 7).

3. The initial justification for airport expansion offered by the city, the county, and community leaders was to make Louisville competitive as an airline hub by building parallel runways. Later, it came out that the real purpose of the expansion was to serve UPS's expansion goals and create jobs. Approximately 3,500 residents were relocated in three neighborhoods to make room for the expansion. The City of Louisville initially used its powers of eminent domain to purchase the houses of these residents, but this later was ruled unconstitutional by the state supreme court as a violation of the state statute authorizing eminent domain. Although eminent domain was authorized by the state statute for urban renewal to address urban blight, the city's rationale that noise from interstate highways and other sources constituted blight in three neighborhoods was considered by the court to go beyond the language of the statute. By this time, however, so many homes had been purchased that neighborhood leaders accepted a settlement offer from the city that resulted in the three neighborhoods being taken for the expansion. The UPS expansion was expected to create about 5,500 jobs.

4. A committee headed by Robert Taylor, dean of the College of Business at the University of Louisville, cataloged "problems expressed with current organizational arrangement for Economic Development: business leaders and businesses confused as to who does what in economic development; widespread concern about the number of organizations that seem to have similar functions and compete for limited public and private monies; no overall, coordinated community strategic plan for economic development; lack of effective leadership organization that deals with the long-term community agenda; lack of effective, active regional business organization; lack of communication and coordination among existing economic development organizations; need to strengthen relationships with state economic development officials; perception that business leadership gives more attention to business attraction than to support for existing businesses where significant job growth occurs; perception that some services provided by local government would be more appropriately provided by private sector" (internal communication titled *Plan for Organizing Economic Development in the Greater Louisville Region,* 1990, p. 3).

5. An analysis of the project participants (Chilton, 1993) raises questions about how reflective the 187 committee members were of regional interests and whether the project really represented diversity within the community. Chilton finds that the project did expand inclusion to southern Indiana (about 16% of project participants) and African Americans (about 14% of project participants). The project did less well, however, at involving women (only 24% of participants). In terms of geographic representation, Jefferson County was overrepresented, with approximately 75% of the participants (above its 70% share of the region). Southern Indiana representatives made up about 16% of the participants (about even with its 19% share of the region). Only 3.2% of the participants were from Shelby, Oldham, and Bullitt Counties combined (compared with those counties' 11% share of the regional population). The CBD was overrepresented, with about 41% of the members employed there. Louisville's South End (home to a large blue-collar population) and West End (home to most of the region's African American population) were underrepresented, with 5 and 3 participants, respectively.

6. Several of the Cornerstone 2020 reports are relied on for population and job projections used in this chapter.

7. There are six classifications of cities in Kentucky, based on population size. A city's classification (first through sixth) dictates its functions and powers and the type of city government it may operate under (council-manager, mayor-council, commission).

8. Occupational taxes are "license fees imposed upon the 'privilege' of working and/or engaging within the City of Louisville and Jefferson County in a business, profession, trade or occupation" (Sinking Fund, pamphlet, undated). In essence, occupational taxes are a proportional tax on "earned" income. They are levied by place of work.

9. The agreement was hammered out in December of 1985, was passed by the state legislature in early 1985, was adopted as local ordinance in the summer of 1985, and took effect in January of 1986.

10. Actually, functional consolidation in the area of economic development was achieved in the prior year, when the city and the county each appointed the same official to head their respective offices (see Vogel, 1990).

11. There was very little opposition to the compact, except on the part of the 93 small cities in Jefferson County, which were unhappy about changes in their ability to annex. The major impact of the compact on these small cities was the elimination of their ability to annex any of the territory that the city of Louisville had included in its annexation efforts. If or when the compact ends, the City of Louisville has the right to move on these areas. Additionally, under the compact, no new city may be incorporated. The small cities and some residents of unincorporated areas of the county who wish to incorporate new cities do not accept this interpretation of the compact, questioning how an agreement they are not party to can bind them.

Besides the concerns raised by small cities, the League of Women Voters expressed some dismay over how fast the compact was adopted, having been proposed in December of 1985, then passed by the state legislature in March of 1986, then passed by the city and county governments in June, and taking effect in July of that same year, with little opportunity for community input or debate. The League feared that politics and patronage would intrude into the agencies' operations and that there would be less opportunity for citizen participation in decisions. The League has monitored the compact's effects on the independent agencies and found that services have not been affected significantly.

12. Not everyone believes that the airport expansion project should be viewed as an economic development success, and questions have been raised about the decision-making process leading to the expansion (see Vogel, 1990).

13. According to one county official, when one examines who the recipients of county services are (particularly jails and health), city residents receive more of these benefits than do county residents. The city would respond that city residents are also county residents and are paying for the services anyway.

14. The change in the county's perspective on the compact is dramatic. While Harvey Sloane was still judge, Jefferson County made a presentation at the KACO annual convention in 1987 titled "Louisville/Jefferson County 'Compact': The Alternative That Works."

15. 1992 Sub-county population estimates of the U.S. census report a .7% growth rate for the city and a .9% growth rate for the county between 1990 and 1992, reversing a 40-year trend in the city and a 20-year trend in the county.

16. A special session of the legislature in the spring of 1995 approved state aid for the stadium. This aid takes the form of a swap of state property to the owner of the land where the stadium is to be built. In addition, the legislature agreed to provide funding toward the cost of expanding the downtown convention center.

17. The Fiscal Court member was Darryl Owens, an African American who led the antimerger forces in 1982 and 1983.

18. The steering committee facilitator was prohibited by the steering committee from participating in a forum on merger sponsored by a community group in January 1995.

19. The authors, H. V. Savitch and Ronald K. Vogel, were hired by the Jefferson County government in January 1996 to analyze the recommendations of the citizens' task force.

20. Some members of the African American community believe that the push for local government reorganization by business leaders is a result of concern about "tipping," or that African Americans eventually will dominate the central city if present population trends continue.

21. These criteria to judge the effectiveness of metropolitan governance are drawn from Barlow (1991).

22. The Metropolitan Sewer District (MSD) and Transit Authority of River City (TARC) have their own revenue sources. MSD charges fees, and TARC relies on fees and an occupational tax. Both, to some extent, have the ability to operate beyond the county borders.

REFERENCES

Adams, J., & Bartlett, B. (1995, October 1). Racial face lift alters suburbia. *Courier Journal,* p. 1A.

Angotti, T. (1993). *Metropolis 2000.* Routledge: London.

Barlow, I. M. (1991). *Metropolitan government.* Routledge: London.

Chilton, K. M. (1993, April). *Growth machine theory and its applicability to Louisville: An economic development case study.* Paper presented at the Urban Affairs Association Conference, Indianapolis.

Copas, K. (1995, October 2). Louisville and southern Indiana: Competition and cooperation. *Business First,* p. 25.

Cornerstone 2020. (1994). *Greater Louisville forecasts of jobs, population, and income: 1995 to 2020—final report.* Louisville: Author.

Cornerstone 2020. (1995). *Jefferson County forecasts of people, jobs, and housing: 1995 to 2020—draft report.* Louisville: Author.

Ehrenhalt, A. (1989). For chambers of commerce and cities, the days of conflict may be over. *Governing, 3*(2), 40-48.

Finance, Compact, and Governmental Structure Study Group. (1995, October). *Working paper.* Louisville: Jefferson County Governance Project.

Forecast Louisville. (1987). *Forecast Louisville: A community agenda for the Louisville economic community.* Louisville: Louisville Area Chamber of Commerce.

Kleinberg, B. (1995). *Urban America in transformation.* Thousand Oaks, CA: Sage.

Louisville/Jefferson County Office for Economic Development. (1990). Louisville enterprise zone first in U.S. to reach $1 billion in capital investment. *Business Leader, 4*(1).

McDonough, R. (1994, October 12). Grafton is quitting Jefferson County governance panel. *Courier Journal,* p. B1.

National Academy of Public Administration. (1980). *Metropolitan governance: A handbook for local government study commissions.* Washington, DC: U.S. Department of Housing and Urban Development.

Negrey, C., & Vogel, R. (1991, April). *Policy indicators for economic development.* Paper presented at the Urban Affairs Association, Vancouver.

Price, M., Coomes, P., & Hoyt, W. (1995, September). *Occupational tax collections in Jefferson County place of work vs. place of residence.* Paper presented to the Jefferson County Governance Project.

Raymond, L. (1992, November 2). Big gain in jobs forecast for area. *Courier Journal,* p. B7.

Regional Economic Development Strategy. (1992). *Implementation plan, final report.* Louisville: Louisville Area Chamber of Commerce.

Regional Economic Development Strategy. (1993). *Investment strategies, 1993-1997.* Louisville: Louisville Area Chamber of Commerce.

Vogel, R. (1990). The local regime and economic development. *Economic Development Quarterly, 4,* 101-112.

Vogel, R. (1994). *Local government reorganization.* Louisville: League of Women Voters and Department of Political Science.

7

Pittsburgh: Partnerships in a Regional City

LOUISE JEZIERSKI

Fifty years of effort by one of the country's most formidable public-private partnerships has created a new Pittsburgh. An urban renaissance has been achieved, recognized by Rand McNally when it rated Pittsburgh the "Most Livable City" in the United States in 1985. Since the 1940s, corporate leadership in Pittsburgh has worked with city, county, and state governments to implement numerous plans that have re-created the area's physical and employment structure. Pittsburgh was once a mighty manufacturing region of machinery and primary metals. It is now a "postindustrial" economy, centered on "specialized, advanced services" catering to the needs of a headquarters city and expanding into high technology sectors of software, engineering, medicine, and education. This was accomplished during three periods of clustered development projects labeled Renaissance I (1943-1970), Renaissance II (1978-1988), and Renaissance III (1982-present) (Stewman & Tarr, 1982; Jezierski, 1988).

The restructuring of the regional economy turned the "Steel City" into "Software City," but it also cost jobs and set communities adrift. The success of a concerted regional effort to attempt economic transition must be considered in the context of massive decline in manufacturing. The postindustrial transformation of the metropolitan economy bolstered the city center while many industrial suburbs became deracinated. Redevelopment has been uneven, and the region has not been able to sustain population growth (see Table 7.1).

Pittsburgh is more fragmented than any other metropolitan area. The economic influence of the Pittsburgh economy extends along the Ohio River Valley to Ohio and West Virginia. This region is defined by its

159

TABLE 7.1 Population Trends for Pittsburgh and Its Metropolitan Area

	Pittsburgh	Allegheny County	Metropolitan Area		Ratio of City to County Population
1950	676,068	1,515,237	2,213,136		.45
1960	604,332	1,628,587	2,405,435	SMSA	.37
1970	520,089	1,517,996	2,401,362	SMSA	.34
1980	423,938	1,450,085	2,218,870	SMSA	.29
1990	369,879	1,336,449	2,056,705	PMSA	.28
Percentage changes					
1950-1990	−45.3	−11.8	−7.1		
1960-1990	−38.8	−17.9	−14.5		
1970-1980	−18.5	−4.5	−7.6		
1980-1990	−12.8	−7.8	−7.3		

SOURCE: *County and City Data Book*, U.S. Bureau of the Census, 1952, 1962, 1972, 1983, and 1992.

rugged topography of river valley towns and hollows, ravines and mountains, and the extensive bridge and highway systems that are required to connect them (Figure 7.1). The 1990 U.S. Census defines the Pittsburgh CMSA as including Allegheny County, which contains Pittsburgh, and the contiguous counties of Beaver, Fayette, Washington, and Westmoreland. Allegheny County has 129 municipalities, but with the addition of special districts and authorities, the County incorporates about 323 governmental units (Advisory Commission on Intergovernmental Relations [ACIR], 1992). The City of Pittsburgh contains more than 80 distinct neighborhoods, some wealthy and some poor. The fragmentation and disparity hold in the suburban ring as well. Unlike the typical American metropolitan pattern of suburban boom and center city decline, Pittsburgh provides an unusual case of suburban decline and downtown revival. Central city versus suburban disparity of income indices are rather low, but inequality among cities in the area seems to have increased since 1980 (Nathan & Adams, 1989; Miller, Miranda, Roque, & Wilf, 1994). A strong interdependence between the region and the City of Pittsburgh is evident because the city center contains 40% of area jobs, but the political economy of the city differs from that of the region. The City of Pittsburgh has succeeded in a transition to a postindustrial economy, based on corporate, medical, and educational services, but the steel valley communities in the suburban fringe still have not recovered from the deindustrialization crises of the 1980s.

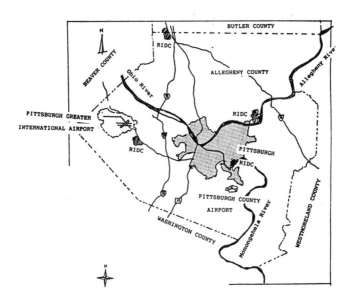

Figure 7.1. The Pittsburgh Region

The uneven and fragmented structure of the region belies the considerable centralization of economic decision making found there. The proliferation of governmental units has been interpreted by the "public choice" approach as the result of extensive, market-based niches of collective consumption (ACIR, 1992).[1] This ahistorical view fails to account for the fact that industrial corporate decision makers were, in part, responsible for this fragmentation initially and that they also were early champions of a rationalized and centralized metropolitan government system. Instead, Pittsburgh has forged a "third way" of producing urban regeneration: Nonprofits have become a key engine of economic growth (Sbragia, 1990). Pittsburgh's transition was constructed through the organizational capacity of a public-private partnership. A centralized network of business and government organizations has provided an effective tool for setting development goals and implementing them. The partnership in Pittsburgh has been characterized as corporatist (Coleman, 1983), consisting of centralized, institutionalized, and cooperative relations among key, recognized, legitimate interests from the private, nonprofit, and public sectors. The partnership structure has proved to be durable and flexible, and it has been constructed through negotiation and compromise. The partnership directs and implements development plans through interorganizational cooperation and reliance on social networks, values of paternalistic commitment, and rational planning and management tools.

Public-private partnerships represent new community-based approaches to coordinating resources in an era of post-Keynesian state restructuring. They contribute to the formation of the "entrepreneurial" or "broker" state and "contracting regimes" (Eisinger, 1988; Osborne & Gaebler, 1992; Smith & Lipsky, 1993). I argue that they are characterized by a centralization of decision making. The regime of decision making for the region is centered in downtown Pittsburgh, especially within business-based groups. This form of governance enables local political and economic actors to adjust effectively to global political economic trends, but they are not as efficient or legitimate as some would hope. Thus, challenger groups and social mobilization inevitably emerge in response to lack of incorporation. The organizational capacity of the business sector has required and encouraged the growth of parallel nonprofit organizations such as neighborhood groups and alliances, semiautonomous governmental authorities, and worker-led initiatives. As a result, some new voices become incorporated. A key condition to successful partnerships is that the authority of the state must be reinforced or expanded to implement plans.

■ **A Short History of the Pittsburgh Regional Political Economy: From Gateway to Steel to Corporate Headquarters**

Even an abbreviated history of the Pittsburgh region can provide some clues to the nature of the contemporary metropolitan political economy and culture (Hays, 1989; Muller, 1989; Lubove, 1969). The Gateway city began as Fort Pitt in the late 18th century; its purpose was to oversee the new "Western" frontier. Its strategic location provided control and access to the riverways, lands, and trade downriver, thus supporting an entrepôt site for lumber, shipbuilding, and iron products, along with farming trade. Pittsburgh's privileged trading position was soon eclipsed, however, by railroads, except in the case of steel distribution.

The industrial character of the "Iron City" was established by the Civil War. Area towns grew with the development of indigenous regional resources such as oil, natural gas, and coal, and later with iron, glass, textiles, steel, and machinery production. Production facilities were located in small cities along the rivers, such as the "Mon Valley" steel mill towns along the Monongehela River. Coal mines and iron furnaces scattered throughout the region were integrated into a centralized network of Bessemer works dominated by a downtown Pittsburgh elite of financial and engineering entrepreneurs. The political isolation of manufacturing villages was constructed through the establishment of paternalistic and repressive company towns. The power of the owners situated in the City of Pittsburgh, and built literally into the skyline, helped to create an atmosphere of domination in the industrial neighborhoods and valleys below.

The hilly terrain that concentrated workers' settlements in the hollows contributed to tightly knit communities that integrated work life and ethnic ties and supported heightened workplace and community solidarity (Houston, 1979a, 1979b, p. 82; Bodnar, Simon, & Weber, 1982; Ostreicher, 1989). Some of the country's fiercest labor battles took shape here, but isolation created uneven unionization and divided collective organizing efforts. Work life permeated community life, and town names or neighborhoods became associated with corporations or with specific factories. The geography of industrial organization maintained class distinctions between mill towns and suburbs that still exist.

■ Annexation, Planning, and Consolidation of the Metropolitan Landscape

Concerns over metropolitan planning and regional problem solving appeared as early as 1900. Rationalizing deep ravines, forceful rivers, and steep hillsides required a regional effort for building a network of bridges, canals, turnpikes, roadways, and railways to connect the fragmented area topography. The depth of social and environmental problems in the region, starkly revealed by The Pittsburgh Survey (1907-1908), also required regional solutions. Reform efforts were addressed through a centralized organizational infrastructure.

One obvious solution was annexation. The Pittsburgh Board of Trade suggested annexing its northern rival, Allegheny City, as early as 1854, and this finally was accomplished in 1907, along with annexation of other towns. The drive for a "Greater Pittsburgh" form of government was pushed repeatedly by the Pittsburgh Chamber of Commerce, and concerns heightened when Pittsburgh was eclipsed in the 1920 Census by cities including Cleveland and Detroit. This aggressive mandate was answered by a cooperative and protective arrangement of municipalities organized as the League of Boroughs and Townships of Allegheny County. Metropolitan government was pursued as a compromise by the corporate-led Chamber of Commerce and Board of Trade. A proposal to create the first metropolitan government in the country and a constitutional amendment to create a federated city and county government for Pittsburgh and Allegheny County had state support but was rejected by voters in 1929. Meanwhile, municipal reform in Pittsburgh advanced with the creation of a City Planning Commission and a Department of City Planning in 1911. The Allegheny County Planning Commission was organized in 1923. These agencies, however, were limited to regulatory functions, and planning agendas were still set by the private sector.

The period before World War II provided a foundation for cooperative action within the private sector. Elite voluntary associations proved to be the crucible of regional physical, environmental, and social planning. They supplied a tradition of corporate commitment, a consolidated network of organizations, and a prioritized redevelopment agenda. Groups such as the Pennsylvania Economy League (western division) and the Pittsburgh Regional Planning Association, both organized in the 1930s, served as the primary research arms for both downtown and regional concerns (Lubove, 1969; Stewman & Tarr, 1982, p. 62). These voluntary efforts, however, could not address the pervasive and far-reaching problems facing the region.

By the 1940s, the decline of the city was immediately visible in the thick smog and smoke pollution that permeated the environment: Streetlights literally were kept on during the day. There also was a continuing threat of floods. Twelve major floods occurred between 1900 and 1950, and the flood in March, 1936, devastated the downtown triangle area. The Depression of the 1930s exacerbated the problems of an already faltering steel industry (Houston, 1979a, 1979b). General deterioration of business conditions was evident in a 20-year halt in building construction downtown and the decay of the central business district, a lack in the modernization of industries, traffic problems, and housing shortages (The Pittsburgh Renaissance Project, 1974, p. 27). Labor struggles also threatened the corporate elite. A number of corporations seriously considered relocation.

These crises created a sense of urgency that prompted business to engage local and state government in a redevelopment partnership. The private sector saw its planning efforts as "form without substance"; it could not implement plans without the legal authority provided by the public sector, and this power was relatively weak (Lubove, 1969, p. 87). A more powerful vehicle for tackling area problems was required, so the first item on the redevelopment agenda was increasing state power.

■ Partnership and Renaissance I: 1943-1970

Partnership consolidation dates from the incorporation of the Allegheny Conference on Community Development (ACCD) in 1943 and the election of Mayor David Lawrence (Democrat) in 1946. "The Conference," as it is known, was organized by Richard King Mellon and includes a select membership of local corporate chief executives.[2] A small staff, headed by an executive director, coordinates the group's efforts. The ACCD made plans for a loose set of individual projects aimed at improving the downtown area and that resulted in a reconstruction of the regional economy. The culmination of these projects was later referred to as the Pittsburgh Renaissance. The organizational capacity created during this period provided an engine that continued Pittsburgh's revitalization over several decades.

Redevelopment under Renaissance I was initiated by the private sector with a division of labor among business group nonprofits: The PRPA drew up the plans, the Economy League did research and organized funding, and the ACCD worked with public authorities. First, a consolidation and extension of state authority had to come about, just as the fragmented private sector had to become unified before it could act. It could be argued that the public sector's redevelopment efforts were organized in response to the private sector. The city had neither the technical staff nor the funds to develop renewal plans before the creation of the Urban Redevelopment Authority (URA) in 1946. The URA provided the legal means to coordinate and expedite land use coordination through use of eminent domain and federal development money.

The Pittsburgh Renaissance was first a legal matter; its foundation was the passage of bundled proposals in the Pennsylvania legislature in 1946. A nonpartisan, cooperative effort, including the Economy League, the PRPA, the ACCD, Allegheny County, and the City of Pittsburgh developed "The Pittsburgh Package." Eight of the 10 bills were passed, including those related to county smoke control legislation that overrode railroad exemptions; countywide refuse disposal; expansion of county planning commission control over suburban subdivision plans; the creation of a Pittsburgh parking authority, a county transit and traffic study commission, and a Department of Parks and Recreation; the expedition of the Penn-Lincoln Parkway (highway); and broadening the Pittsburgh tax base to include sources other than real estate. Another important piece of legislation, passed in 1947, allowed insurance companies to invest in redevelopment areas (Stewman & Tarr, 1982). A state redevelopment law was passed in 1945 that allowed municipalities to establish redevelopment authorities to acquire and clear land certified as blighted and to raise money through bonds. Mayor David Lawrence also used his clout in county politics to influence the state legislature. Lawrence became governor of Pennsylvania (1959-1963) and provided continued ties to state-level political power, which was necessary for implementing roadway and bridge reconstruction, the $100 million flood control projects, $300 million in airport construction, and of course, the public authorities that coordinated development and financing. Thus, much of the Pittsburgh renaissance was metropolitan in scope from its inception, and the creation of extralocal public authorities was a crucial implementation tool.

The expressed goal of the Improvement Program that became Renaissance I was to create a more attractive physical environment to keep corporate headquarters from leaving. Revitalization strategies used during the Renaissance I era were first environmental, including smoke abatement and flood control, followed by physical development. In addition, industrial, commercial, transportation, and entertainment investment was targeted. Redevelopment efforts resulted in refashioning the function of the downtown area. The URA condemned physical structures through eminent domain that previously were warehouses, port facilities, factories and workshops, and rail entrepôts. The once dense, industrial zone located at the intersection of Pittsburgh's three rivers was enhanced by high-rise office buildings, hotels, retail shops, luxury apartments, and Point State Park. This would become known as the "Golden Triangle." Renaissance I enhanced the infrastructure for administrative service functions for the local industrial and financial corporations that dated from the late 1800s.

Studies of the regional economy were key to shaping development plans, notably the *Economic Study of the Pittsburgh Region* undertaken by the PRPA and published in 1963. This report stressed the need for greater economic diversity and further investment in technology and R&D to reduce dependency on manufacturing (Lubove, 1969, p. 135). Although

Renaissance I was primarily a "bricks and mortar" effort focused on downtown development, the program did not ignore the traditional, regional industrial sector, especially in reinforcing the steel industry. Postwar investment (upward of $1 billion) boosted industrial production in the steel valleys well into the 1970s, even though the industry already was losing some of its competitive edge (Houston, 1979a, 1979b; Hoerr, 1988).

The ACCD created spin-off organizations designed to promote job development directly. In 1955, the Conference created the Regional Industrial Development Corporation (RIDC) to set up industrial parks. The RIDC works as a regional economic development corporation and assists companies with site selection, facility design and construction, and the packaging of low-cost financing from local, state, and federal incentive programs. It has the power to buy, sell, and finance buildings. Three industrial parks begun by the RIDC in the Pittsburgh environs include a 600 acre facility in northern Allegheny County established in 1964, the 925 acre Thorn Hill facility in Butler County established in 1971, and a 500 acre facility established in 1979 to the west, near the airport, that specializes in international commerce and serves subsidiaries of foreign companies. All are located within 15 miles of the Golden Triangle. The three parks include more than 130 buildings and employ about 15,000 workers. The fourth park opened in 1988 at the former site of a closed Jones and Laughlin steel mill in Pittsburgh's South Oakland neighborhood and serves as a high-tech industrial park. The RIDC also is the primary developer of incubator facilities, with two located in the Oakland/university area known as University Technology Development Centers (UTDC).

Penn's Southwest, created in 1971 by the ACCD, is an organization that coordinates efforts to attract firms to Pittsburgh, mostly branch offices. It is estimated that between 1976 and 1986, more than 300 firms and 38,000 new jobs were gained in the Pittsburgh area as a result of this organization's efforts, although less than half of these were in manufacturing. It has a global mission, with offices in Europe and Tokyo. It successfully attracted a Volkswagen plant (now closed), Bayer AG's U.S. headquarters, and Mobay Chemical and a number of Japanese branch plants. In 1980, Penn's Southwest decided to go after high technology and added offices on the West Coast to follow high-tech as well as Pacific Rim business. This strategy also brought an unexpected but welcome boost to the area food industry.

■ Community Mobilization and
the Interlude (1970-1977)

When the partnership began to redevelop the downtown area under Renaissance I, it was not challenged because the community perceived the city center as the ACCD's "own turf." Other projects, such as mass transit and metropolitanization, were rejected. Opposition was strongest when community groups mobilized in defense of their neighborhoods. An impasse was created between the partnership and community groups, which organized alternative development schemes. In time, the demands and tactics of neighborhoods escalated to demanding involvement in the partnership itself. The targeting of federal money to community-based nonprofit organizations in the 1960s and the advent of the civil rights movement allowed neighborhood groups to establish an independent political base. The response of the public-private partnership has been to become increasingly involved with neighborhood development, but it has been a 40-year lesson (Jezierski, 1990).

The downtown agenda of the public-private partnership was rejected by Pittsburghers when they elected Pete Flaherty as mayor (1969-1977). He ran without the support of the Democratic Party, labor, or the business community. His populist administration made key changes in development policy that widened neighborhood participation. As a result, the earlier consensus between the mayor's office and the Allegheny Conference broke down and new downtown investment plans came to a halt. The business community saw the Flaherty years as chaotic, angry, and counterproductive, although it acknowledged his efforts to streamline government. This period of breakdown in the partnership's history was summarized by Stewman and Tarr (1982) as the "Interlude."

While Flaherty dissolved the city's partnership relation with the private sector, a countywide partnership was forged. The three county commissioners accepted the help of the Chamber of Commerce and Mellon Bank and formed the Committee for Progress in Allegheny County (ComPAC). Its board was created from chief executive officers of Allegheny County corporations, and nine task forces were organized to provide management reforms. Reorganization of the county administration produced $3 million in one-time savings and another $3 million in annual savings (Stewman & Tarr, 1982, pp. 85-86). Intergovernmental cooperation was established with the Authority for Improvements in Municipalities, organized through Allegheny County's Department of Development, for infrastructure improvements and maintenance.

■ Manufacturing Declines and the Regional Economy Restructures

Manufacturing employment decreased drastically in the Pittsburgh region (for 10 counties of southwestern Pennsylvania), from 269,000 in 1980 to 183,000 in 1983. Production workers in the city of Pittsburgh decreased 76% between 1947 and 1982.[3] Mellon Bank estimates that the region lost 60,000 jobs in primary metals, plus another 50,000 in manufacturing, between 1980 and 1986. While manufacturing decreased 6% nationally during this period, Pittsburgh experienced a 44% loss of manufacturing jobs (Hathaway, 1993, p. 27). Manufacturing accounted for almost one-third of all jobs in the region in 1970, but only about 14% of the region's labor force worked in manufacturing by 1990 (compared to 19% nationwide). Steel employed more than 80,000 workers in the 1940s and 35,000 in 1981, but by 1987 only 4,000 jobs were left in "steel city" (Hathaway, 1993, p. 26; Jacobson, 1987).

Long-term unemployment followed this swift and brutal restructuring. The unemployment rate for the city in 1982 was 10%; it was 16% regionally. In some Monongehela Valley towns, unemployment reached 50%, threatening the fiscal stability of suburban cities. In a survey undertaken by the *Pittsburgh Post-Gazette* in 1985, only 59% of unemployed factory workers in the area had found new jobs, and many got jobs that paid much less than what they earned in the mill. The immediate needs of these workers, and the mill towns and neighborhoods in which they lived, overwhelmed social welfare programs for unemployment compensation, welfare benefits, and social security. Food banks, mortgage pools, and suicide and substance abuse counseling were organized by the displaced workers. Many unemployed workers and their families moved out of the region. The metropolitan area lost more than 7% of its population between 1980 and 1990 (see Table 7.1) and may help account for the low unemployment rate for the region (5%) by 1990.

Over the long term, the area economy made compensatory gains in service industry jobs (see Table 7.2). By 1990, metropolitan employment in services accounted for more than 28% of the labor force. For the City of Pittsburgh, almost 40% are now employed in service-related sectors, and the unemployment rate was as low as 4.5% in 1990. More were employed in high-tech jobs than in steel by the mid-1980s. Between 1978 and 1988, the six major educational institutions increased employment by 26% (mostly at the University of Pittsburgh and Carnegie-Mellon University). In 1988, 11,000 people worked for the top seven industrial firms,

TABLE 7.2 Distribution of Employment by Industry (in percentages) in the Pittsburgh Region

	1970	1980	1991
Construction	5.3	5.3	5.3
Manufacturing	31.7	25.6	13.8
Nondurables	5.8	5.0	4.0
Durables	25.9	20.6	9.7
Transportation, communication, and public utilities	7.1	7.7	6.7
Wholesale and retail trade	20.4	21.3	19.9
Finance, insurance, and real estate	4.5	5.4	6.9
Services[a]	25.4	27.0	28.4
Public administration (government)	4.1	3.4	11.9
Total number of workers	870,902	938,473	1,044,000

SOURCE: Data for 1970 and 1980 are from *General Social and Economic Characteristics of the Population*, U.S. Bureau of the Census. For 1970, the data are from Table 87; for 1980, data are from Table 122. Data for 1991 are from *Geographical Profile of Employment and Unemployment for 1991* (Table 26), Bureau of Labor Statistics, 1992.
NOTE: The Pittsburgh SMSA for 1970 and 1980 includes Allegheny, Beaver, Washington, and Westmoreland Counties. The 1991 data are for the Pittsburgh-Beaver Valley CMSA and include Allegheny, Beaver, Fayette, Washington, and Westmoreland Counties.
a. Includes the following categories: business, repair, private household and other personal services, entertainment and recreation, and professional and related services (hospitals, health, elementary and secondary schools, colleges, related educational services, social service and religious organizations, and legal and engineering services).

37,000 worked at universities and colleges, and 23,400 worked in the eight major hospitals (Sbragia, 1990).

Crisis renewed a call to Renaissance. Restructuring and the recession of 1979 ushered in concerns over job losses and community preservation. These issues brought about profound struggles over economic and community development policy. The organizational foundations laid during Renaissance I were revived easily, and although the personal direction of the Mellons is now gone and corporations have changed, the organizational capacity endures. The organizational structure has become more complex, however: Neighborhood interests have become more formal and technical, the private sector has created specialized civic associations, and labor union representation has dwindled. The partnership structure also has been strained by tensions over new demands over participation. The recession of the early 1980s produced a coalition of labor unions, churches, and neighborhood groups mobilized against economic disinvestment. Protests over plant closings and proposals for takeovers by worker-management groups were organized by labor unions, especially along the industrial Monongehela River valley. Corporations, civic leaders, and the increasingly prominent nonprofit organizations moved to forge a "new Pittsburgh" based on services and software.

■ Renaissance II: Downtown and Neighborhoods (1978-1988)

Under the independent Democratic administration of Richard Caliguiri (1977-1988), a renewed commitment to economic development and to the partnership was set in motion, a process he referred to as Renaissance II. On his first day in office, Caliguiri called the Allegheny Conference and reestablished a working relationship with it. He also encouraged ties between neighborhood groups and the ACCD and foundations. He emphasized consensus and repudiated the conflict style of his predecessor, Mayor Flaherty, and of activist neighborhood organizations. The dualistic approach that Caliguiri pursued continued downtown development, adding new hotels, retail businesses, riverfront development, and a new subway system, as well as promoting neighborhood renewal. Airport expansion also was organized with the Pennsylvania Economy League. After Caliguiri's death, Mayor Sophie Masloff (1988-1993) continued the strong relationship between City Hall and neighborhoods, but economic development in the city and region were left to flow from the initiatives organized in the early 1980s. In late 1993, another mayor from activist neighborhood roots, Tom Murphy, was elected.

The major thrust of Renaissance II was to bolster headquarters activity. This was important once again because of corporate restructuring and consolidation, as well as the more usual, cyclical market needs for office space. Corporate mergers and local plant closings threatened the corporate administrative sector of the Pittsburgh economy in the early 1980s. Moreover, the loss of large corporate headquarters and production facilities altered the organization of private-sector leadership. Mergers, less familial control of local corporations, the increasing influence of the University of Pittsburgh and Carnegie-Mellon University, and the emergence of new service and high-technology companies within the Pitts-burgh economy have changed participation in the partnership.

The recession of the early 1980s surprised the corporate community, and many business leaders said that they did not expect the huge loss of manufacturing jobs, although many suspected that automation would take away production jobs over the long term. They had hoped that diversifi-cation into light industry, pursued through RIDC industrial park develop-ment, would alleviate these problems. The massive loss of jobs brought a swift transformation that left gaps in the economy overall. These events led the partnership to reassess regional economic development and map out a long-term economic planning effort that recognized the need to build whole new sectors. Small-scale, entrepreneurial efforts and high technol-ogy were targeted. Thus, a new initiative toward regional economic diversity extended the partnership into what may be called Renaissance III, which emphasized technology and services. These areas were seen as indigenous but underutilized remaining strengths in manufacturing re-search and development that could be advanced.

■ Renaissance III: High Technology and Services (1982 to the Present)

The Allegheny Conference organized an Economic Development Committee in 1981 to better understand the regional economy. The final report, "A Strategy for Growth: An Economic Development Program for the Pittsburgh Region," issued in 1984, provided the rationale for an economic development strategy outlined in the following five points.

1. The forces that caused Pittsburgh's decline are irreversible, and the area will never return to primary metals and durable goods manufacturing. Nor will Pittsburgh be allowed to be so vulnerable to a single type of industry in the future. The economy must diversify to enjoy continued stability.

2. A development strategy should pursue a mix of business in different industries, both product- and service-oriented industries, and a mixture of mature and new activities.

3. The strategy should be long term and avoid quick fixes. There is no way to solve the massive unemployment problem within a year or two, but a long-term solution of developing employment sources is less likely to have severe cycles of unemployment in the future.

4. The strategy should be oriented to the private sector and market driven, and it should encourage private investment. A strong and supportive public sector is important, however, to provide transportation improvements, training and retraining, tax policy, and a supportive climate for investment.

5. The strategy should call for coordinated action rather than central planning.

Once again, implementation of the Economic Development Committee's recommendations would require state funding and increased authority capabilities. A bundled package of proposals called Strategy 21 (Office of the Mayor, City of Pittsburgh, Pennsylvania, 1985) was prepared by the ACCD in cooperation with city and county officials for presentation to the Pennsylvania state legislature. Spending requests included $97 million for improvements at the Pittsburgh Airport; $57.5 million for three Pittsburgh redevelopment projects, including the Three Rivers Stadium project, the Strip District project, and the Herr's Island project; $49 million for 11 redevelopment projects in the Monongehela Valley; $150 million for highway improvements, including four Monongehela Valley highways; and $71 million for advanced technology research at the universities in Pittsburgh.

"Strategy 21" was controversial. The Partnership professed that its goals were designed in the public's interest, yet all the budget requests were for large capital improvements that did not directly address the needs of neighborhoods or labor. The partnership saw small business development and coordinated retraining as the key to structural unemployment. Neighborhood groups and Monongehela Valley communities felt that the indigenous, skilled labor force and labor unions were written off. Moreover, these traditional constituencies felt shut out of the decision-making process because development planning relied on coordinated private-sector initiatives.

The new hope and strategy for economic revitalization is the promise of high-technology growth. The ACCD and Penn's Southwest assessed the area's resources and decided that pursuing high tech was a viable economic strategy. According to the RIDC, the Pittsburgh area boasted of more than 25,000 resident scientists, engineers, and technicians, 170 research laboratories, and a combined corporate budget of more than $1.5 billion spent annually on research and development. By 1984, the southwestern Pennsylvania area employed at least 40,000 people in more than 400 high-tech firms, surpassing the area's remaining 31,000 steelworkers (Lansner, 1985). The technical industry had been evolving in Pittsburgh since World War II, with its foundations in the manufacturing core. The Pittsburgh region claimed more engineers per capita than any other, employed by firms such as USX (US Steel), Alcoa, PPG, and Westinghouse, which have their corporate headquarters as well as research and development centers located there. They have developed new products such as optics and precise instrumentation and control, robotics, and industrial automated controls (Colker, 1985, p. 7). Efforts to attract firms to Pittsburgh, or "chip chasing," were eschewed in favor of a specialized strategy to build a local, industrial "farm system," drawing on inherent regional strengths in software and automation.

Quickly, an organizational infrastructure was built (Ahlbrandt & Weaver, 1987). Penn's Southwest spun off the High Technology Council in 1983. Its board includes representatives from Penn's Southwest, the RIDC, area universities, and large and small corporations. By 1985, it could boast 300 organizational members, representing 131 high-technology companies, 136 professional service companies, and 24 civic, academic, and entrepreneurial research organizations. Universities created their own group, the Western Pennsylvania Advanced Technology Center, in 1983. It focuses on the development of robotics, biological and biomedical technology, high-tech materials, and coal processing. Fostering linkages between small, entrepreneurial companies, universities, and larger, industrial corporations is aided by the Enterprise Corporation of Pittsburgh (1983), which provides management advice. Finally, resources for venture capital were pooled.

Because university research in high technology, medicine, and information is key to new industry development, linkages with universities are essential. The High Technology Council is located in an RIDC building near Carnegie-Mellon University (CMU) and the University of Pittsburgh. When Gulf Oil was acquired by Chevron, its major R&D facility, the Harmarville Labs, was given to the University of Pittsburgh. The Labs have a tremendous capacity, containing 850,000 square feet, 1,000 offices, 62 conference rooms, ready-to-use labs, and a computer center. In 1982, the University of Pittsburgh also began the Foundation for Applied Science and Technology (FAST), a nonprofit research organization to promote collaboration between the university and industry. The University of Pittsburgh, now the city's largest employer, has provided both private- and public-sector organizations with demographic and economic analysis for the region, as well as an adequate supply of highly skilled labor that has helped Pittsburgh's new industries to grow. Also nearby, the Hazelwood J&L steel plant (in the City of Pittsburgh) was razed, and a high-technology incubator, the Pittsburgh Technology Center, which is managed by the RIDC, was built there. The Software Engineering Institute (SEI) was created in 1984 in conjunction with CMU; it soon won a $103 million federal contract to supply defense software.

The partnership of Renaissance III embodies the organizational capacity to develop high technology by coordinating the efforts of small business, universities, local and state government, the Allegheny Conference, and the intermediate organizations of Penn's Southwest, the RIDC, and the High Technology Council. The privatist and corporate character of the partnership was altered to make room for a leadership role for local universities, as well as to promote entrepreneurialism, rather than "managerialism." The tradition of partnership has been adapted and extended to nurture a new industrial development path.

Despite the implication of these efforts for future success, regional unemployment remained the most pressing problem in the 1980s. The devastation caused by the recession of 1979-1984 made the Conference take a second look at the impact of so many unemployed blue-collar workers and their communities. One immediate solution for the fiscal plight of municipalities in the region was an attempt to pool resources through governmental restructuring. The Allegheny League of Cities, the ACCD, and the Greater Pittsburgh Chamber of Commerce established the Intergovernmental Cooperation Program (ICP) in 1982. It helped to set up a number of Council of Governments (COGs) that share a variety of services such as an information and dispatch network, the purchase of public works equipment, a computerized payroll and accounts payable system, street cleaning, solid waste disposal, and administrative training. There are now eight COGs in Allegheny County, serving 114 out of 130 municipalities and 86% of the county population outside Pittsburgh (ACIR, 1992). This arrangement consolidates municipal administrative functions countywide but stops short of creating a "metropolitan government" structure.

Economic development for suburban communities is a long-term issue, and in its latest economic development program, the Allegheny Conference has put more emphasis on job development, retraining, and "improvement of the labor climate." The Education Fund program works directly with schools in the industrial valley towns hardest hit by plant closings. In 1987, the ACCD established the Mon Valley Initiative (MVI) to coordinate putting foundation money into community development corporations in the valley, helping them with housing and neighborhood development (Beauregard, Lawless, & Deitrick, 1992).

The Partnership had decided long ago that manufacturing, especially as it was organized in large, unionized plants, is a thing of the past and not an answer for the future. This perspective ignores the fact that working at the steel plants was a way of life and that workers have had little say in planning their economic future. The stunning swiftness of plant closings, and even dynamiting of factories, left unions, workers, and communities unprepared. Some workers became increasingly radicalized outside of union structures. Mobilization was led by young, activist clergy from local Lutheran and Episcopalian denominations, especially the Denominational Ministry Strategy (DMS). They organized food banks, family counseling, and even job training and employment services. Their first effort was to lobby Governor Thornburgh to declare the Monongehela Valley an economic disaster area, but this was rejected. When Mellon Bank's role in the foreclosure of Mesta Machine was revealed, they moved to more radical tactics, confronting elites at their homes and churches. The DMS (which later changed its name to DMX when US Steel became USX) succeeded in highlighting corporate actions of disinvestment and providing alternative forms of protest. Although this social movement succeeded in changing ideology, it failed to organize workers and community residents into an effective political force (Plotkin & Scheuerman, 1987; Hathaway, 1993).

Workers demanded participation in regional development decision making on issues that affected their livelihoods and communities. They created new alliances to lobby for a different economic development agenda from the one being considered by the Allegheny Conference and the High Technology Council. High technology often seems irrelevant to the skills and desires of the local labor force (Massey, 1985; Kutay, 1989; Fitzgerald, 1990). Neighborhood groups and labor have been pushing for small business development within neighborhood commercial zones, for eminent domain to take over abandoned factories, and for plant closing notification laws.

The most advanced alternative development path was achieved by the creation of the Steel Valley Authority (SVA) in 1985. The SVA is an industrial redevelopment authority incorporating nine participating municipalities.[4] The idea originated with the Tri-State Conference, a group of community and labor activists concerned with steel plant closures, who proposed that communities and workers use the Municipal Authority Act to take over still-profitable plants. The SVA has the power of eminent domain over industrial property and can raise money through bonds and other sources. The SVA could not, however, acquire any steel plants because corporations would not sell to potential competitors. Moreover, the United Steelworkers union gave only guarded support to the idea.

The participation of the City of Pittsburgh was crucial to the success of the SVA plan because city leadership organized technical, personnel, political, and monetary support, including $50,000 that paid for staff. Pittsburgh needed to support its suburban communities, which were suffering from fiscal problems and could not afford to contribute. Tri-State also succeeded in establishing the Pennsylvania Industrial Development Finance Corporation in 1987, a capital fund to be used by local public authorities to address regional industrial disinvestment.

The SVA's most successful effort was creating the largest ESOP (Employee Stock Ownership Plan) start-up company in the United States: City Pride Bakery in Pittsburgh. The ESOP took over a plant owned by Ralston-Purina in a communitywide effort that had the backing of Mayor Masloff, local banks, national and local co-ops, neighborhood groups and churches, and venture capital. The bakery saved 150 jobs and began delivering bread to local supermarkets in Fall of 1992 (Hathaway, 1993).

The SVA's efforts showed that industrial rejuvenation was still a feasible strategy. Moreover, it offered a means of worker and community control over industrial development. The SVA also is an important resource to small communities that need industrial planning expertise. The SVA is a regional solution to economic development that combines resources from various municipalities. Similar to the capacity of COGS to coordinate service delivery, this approach extends intergovernmental cooperation to community-centered economic development. One limitation, however, remains the lack of capital to buy and start up new companies. This alternative development path suggests that communities and nonprofits can coordinate resources to pursue for-profit enterprise.

■ Growing New Regional Economies: Partnerships in a Global Economy

The Pittsburgh case shows clearly that economic revitalization is a political process. Fashioning a postindustrial economy requires creating an organizational capacity that can provide a "re-vision" of the region and coordinate resources through political, corporate, and nonprofit mobilization and organization. The political economic history of Pittsburgh illustrates that "economies are actively constructed, not passively evolved" (Storper, 1985, p. 279).

New technologies and service industries are not developed in a void, however; they are directly linked to the indigenous manufacturing structure of the region. The growth in services is a result of production itself becoming more service intensive (Noyelle, 1983, p. 288). Industrial corporations have served as a primary resource for the development of new industries, through direct innovation in research and development of the manufacturing process, and through their demand for business services such as finance and law. Advanced technology production must be supported by universities, corporate research and development, telecommunications networks, and specialized business services. Pittsburgh has long been a headquarters center, and much of the new and expanded growth there is dependent on headquarters facilities. Pittsburgh once followed only New York and Chicago in Fortune 500 company headquarters but is now fifth. Because most of the area's service jobs are linked to manufacturing industries, these jobs are vulnerable to restructuring as well, and even high-tech jobs were lost in the 1979-1984 recession.

Despite the abundance and wealth of corporate and research resources in Pittsburgh, economic restructuring also required political restructuring. The planning and implementation of a postindustrial development strategy succeeded only with a centralized and durable partnership organization that required building an organizational infrastructure. These conditions are hard to replicate. First, the particular and compelling leadership of the Mellon interests was crucial in organizing the private sector. Second, the corporate resources available in Pittsburgh are shared by only a few metropolitan areas. Third, local government had to extend decision-making authority to private-sector groups that could underwrite the planning process. Much of the technical expertise and financial support for municipal coordination in COGs, high-tech development corporations, and packaging legislation for Monongehela Valley infrastructure enhancement comes from the Allegheny Conference, the Chamber of Commerce, or their spin-off organizations. The deep commitment of local corporations to expand and diversify the regional economy—through airport improvement plans, state and private grant support for university research, and outright gifts such as the Harmarville Labs—is feasible only because available resources were organized in new ways.

The Pittsburgh Renaissance could not have come about without a powerful local government. This was achieved fundamentally through legal means to create semiautonomous authorities; the extension of technical expertise supplied by private research arms such as the PRPA and Economy League, the ACCD, and the SVA, as well as social and economic surveys such as those done by the original Pittsburgh Survey, professional planning departments, and universities; linkages created by interlocking networks that allowed staffs from private and public sectors to coordinate efforts; specialized institutions that coordinated funding resources; and continuity, sustained by both internal and external legitimacy and effectiveness.

This is not to say that the coordination of multiple and fragmented corporate, municipal, state, and citizen interests in regional Pittsburgh has not been fraught with conflict. Struggles over the development agenda have emerged concerning who may participate and the kinds of jobs that should be generated and maintained. These struggles ultimately concern the legitimacy of third-sector, nonprofit governance. The postindustrial transformation in Pittsburgh has not been constructed from mere market mechanisms or some "rational" public choice. Rather, the process has been one of conflict, experimentation, sacrifice and loss, a furious level of organization building, and mobilization of consent.

NOTES

1. The debate over whether a decentralized metropolitan system results in redundancy and requires the reform measures that promote a hierarchical and centralized structure of metropolitan government, or instead whether a decentralized system promotes more "choice," cannot be covered here. For discussion of this debate, see ACIR (1987); Ostrom, Tiebout, and Warren (1961); Bendor (1985); and Miller and colleagues (1994).

2. The Allegheny Conference on Community Development (ACCD) is more powerful than any other private-sector voluntary association, and it made the crucial difference in the establishment and the success of partnership efforts. It is composed of 100 members from the large corporate sector, all chief executive officers. Two rules, strictly guarded, that govern the Conference contribute to its effectiveness. A rule of "no substitution" assigns personal responsibility to members, and no company may directly profit from any project. The "real guts" of the ACCD is the Executive Committee, composed of 24 executives. This committee provides the organization with focus and manageability. The Conference described itself in its 1945 mission statement as "a citizens' group of the Allegheny Region whose purpose is to stimulate and coordinate research and planning, looking to a unified community plan for the region as a whole, and to secure by educational means, public support of projects that are approved by the Conference as part of that overall unified plan" (cited in Stewman & Tarr, 1982, p. 64).

3. U.S. Bureau of the Census, *City and County Data Book, 1947,* Table 5, and *U.S. Census of Manufacturers, 1982,* Table 8.

4. The SVA was incorporated by the Commonwealth of Pennsylvania in November, 1985. The nine participating municipalities are Munhall, Homestead, Turtle Creek, East Pittsburgh, Swissvale, McKeesport, Glassport, Rankin, and the City of Pittsburgh. The Tri-State board believed that these nine represented a strong base because they were diverse

municipalities and they represented diverse industries. Each city has three appointed representatives to a 27-member board. No council member or elected official can be on the board, to avoid conflict of interest.

REFERENCES

Advisory Commission on Intergovernmental Relations. (1987). *The organization of local public economies* (Report no. A-109). Washington, DC: Author.

Advisory Commission on Intergovernmental Relations. (1992). *Metropolitan organization: The Allegheny County case* (Report no. M-181). Washington, DC: Author.

Ahlbrandt, R. S., Jr., & Weaver, C. (1987). Public-private institutions and advanced technology development in southwestern Pennsylvania. *Journal of the American Planning Association, 53,* 449-458.

Beauregard, R., Lawless, P., & Deitrick, S. (1992). Collaborative strategies for reindustrialization: Sheffield and Pittsburgh. *Economic Development Quarterly, 6*(4), 418-430.

Bendor, J. (1985). *Parallel systems: Redundancy in government.* Berkeley: University of California Press.

Bodnar, J., Simon, R., & Weber, M. P. (1982). *Lives of their own: Blacks, Italians, and Poles in Pittsburgh, 1900-1960.* Urbana: University of Illinois Press.

Coleman, M. (1983). *Interest intermediation and local urban development.* Unpublished Ph.D. dissertation, University of Pittsburgh.

Colker, J. (1985). Transformation of an industrial city: High tech comes to Pittsburgh. *Management Review, 74*(May), 48-52.

Eisinger, P. (1988). *The rise of the entrepreneurial state: State and local economic development policy in the United States.* Madison: University of Wisconsin Press.

Fitzgerald, J. (1990). Pittsburgh, Pennsylvania: From steel town to advanced technology center? In R. Bingham & R. Ebert (Eds.), *Economic restructuring of the American Midwest* (pp. 237-254). Boston: Kluwer Academic.

Hathaway, D. (1993). *Can workers have a voice?: The politics of deindustrialization in Pittsburgh.* University Park: The Pennsylvania State University Press.

Hays, S. (1989). Pittsburgh: How typical? In S. Hays (Ed.), *City at the point: Essays on the social history of Pittsburgh* (pp. 385-406). Pittsburgh: University of Pittsburgh Press.

Hoerr, J. (1988). *And the wolf finally came: The decline of the American steel industry.* Pittsburgh: University of Pittsburgh Press.

Houston, D. (1979a). A history of the process of capital accumulation in Pittsburgh: A Marxist interpretation—part I. *The Review of Regional Studies, 9*(1), 12-32.

Houston, D. (1979b). A history of the process of capital accumulation in Pittsburgh: A Marxist interpretation—part II. *The Review of Regional Studies, 9*(3), 81-97.

Jacobson, L. (1987). Labor mobility and structural change in Pittsburgh. *Journal of the American Planning Association, 53,* 438-448.

Jezierski, L. (1988). Political limits to development in two declining cities: Cleveland and Pittsburgh. In M. Wallace and J. Rothschild (Eds.), *Research in politics and society* (Vol. 3, pp. 173-189). Greenwich, CT: JAI.

Jezierski, L. (1990). Neighborhoods and public-private partnerships in Pittsburgh. *Urban Affairs Quarterly, 26*(2), 217-249.

Kutay, A. (1989). Prospects for high technology based economic development in mature industrial regions: Pittsburgh as a case study. *Journal of Urban Affairs, 11*(4), 316-377.

Lansner, J. (1985, April 1). Steel yields in PA. *Datamation,* pp. 22-28.

Lubove, R. (1969). *Twentieth century Pittsburgh.* New York: John Wiley and Sons.

Massey, D. (1985). Which "new technology"? In M. Castells (Ed.), *High technology, space and society* (pp. 302-316). Beverly Hills, CA: Sage.

Miller, D., Miranda, R., Roque, R., & Wilf, C. (1994). *The fiscal organization of metropolitan areas: The Allegheny County case reconsidered.* Paper presented at the North American Institute for Comparative Urban Research Conference, St. Louis.

Muller, E. (1989). Metropolis and region: A framework of enquiry into western Pennsylvania. In S. Hays (Ed.), *City at the point: Essays on the social history of St. Louis* (pp. 181-212). Pittsburgh: University of Pittsburgh Press.

Nathan, R., & Adams, C. (1989). Four perspectives on urban hardship. *Political Science Quarterly, 104*(3), 483-508.

Noyelle, T. (1983). The rise of advanced services: Some implications for economic development in U.S. cities. *Journal of the American Planning Association, 49*(Summer), 280-290.

Office of the Mayor, City of Pittsburgh, Pennsylvania. (1985). *Strategy 21: Pittsburgh/ Allegheny economic development strategy to begin the 21st century: A proposal to the Commonwealth of Pennsylvania.* Pittsburgh: Author.

Osborne, D., & Gaebler, T. (1992). *Reinventing government: How the entrepreneurial spirit is transforming the public sector from schoolhouse to statehouse, City Hall to the Pentagon.* Reading, MA: Addison-Wesley.

Ostreicher, R. (1989). Working-class formation, development, and consciousness in Pittsburgh, 1790-1960. In S. P. Hays (Ed.), *City at the point: Essays on the social history of St. Louis* (pp. 111-150). Pittsburgh: University of Pittsburgh Press.

Ostrom, V., Tiebout, C. M., & Warren, R. (1961). The organization of government in metropolitan areas. *American Political Science Review, 55*(December), 831-842.

Plotkin, A., & Scheuerman, W. (1987, March). *Two roads left: Strategies of resistance to steel plant closings in the Monongehela Valley.* Paper presented at the annual meeting of the Southwestern Social Science Association, Dallas, TX.

Sbragia, A. (1990). Pittsburgh's third way: The nonprofit sector as a key to urban regeneration. In D. Judd & M. Parkinson (Eds.), *Urban affairs annual review: Vol. 37. Leadership and urban regeneration* (pp. 51-68). Newbury Park, CA: Sage.

Smith, S. R., & Lipsky, M. (1993). *Non-profits for hire: The welfare state in the age of contracting.* Cambridge, MA: Harvard University Press.

Stewman, S., & Tarr, J. A. (1982). Four decades of public-private partnerships in Pittsburgh. In R. S. Fosler & R. A. Berger (Eds.), *Public-private partnerships in American cities* (pp. 59-127). Lexington, MA: Lexington Books.

Storper, M. (1985). Technology and spatial reproduction relations: Disequilibrium, interindustry relations and industrial development. In M. Castells (Ed.), *High technology, space and society* (pp. 265-283). Beverly Hills, CA: Sage.

The Pittsburgh Renaissance Project: The Stanton Belfour Oral History Collection. (1974, September). Pittsburgh: Buhl Foundation. (Available at the University of Pittsburgh, Pittsburgh Renaissance Archive)

Part III

Metropolitan Government

8

Miami: Experiences in Regional Government

GENIE STOWERS

Dade County, Florida, the home of Miami, has had a two-tiered regional form of government since 1957. Originally an attempt at experimental metropolitan government (Sofen, 1963) designed to overcome the problems of fragmentation, deal with anticipated growth in the unincorporated areas (Citizens Charter Review Commission, 1986), and an inability to annex to the central city (confidential interview, 1992), the effort has had mixed results. Although the government established in 1957, Metropolitan Dade County, has been widely discussed in the literature on regional governance, true regional governance would involve the entire ecological and economic region in South Florida. Metropolitan Dade County is not yet a regional government—it covers only one county and, some would argue, not even that county very effectively. In fact, the problems in Metropolitan Dade County have arisen as a result not only of the limited nature of its structure in covering Dade County but also of artificially stopping at the Dade County line.

Because of state actions and economic and historical trends, Dade County is a central part of the growing and increasingly interdependent South Florida region. According to the South Florida Regional Planning Council, "Although South Florida is not governed politically as a single entity, the region has commercial and cultural characteristics that transcend political boundaries" (South Florida Regional Planning Council, 1993c). As such, it is the site of such regional governmental institutions as the South Florida Regional Planning Council and the South Florida Water Management District. Aided by a growing sense of "regionalism" and state law, these regional entities form a quasi-regional governing presence, although one struggling to gain power.

185

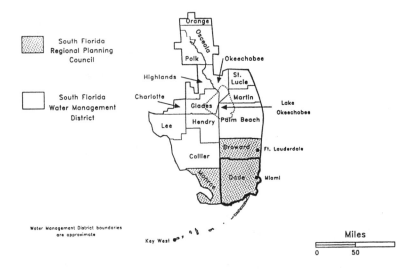

Figure 8.1. Dade County and South Florida

The map in Figure 8.1 indicates the political boundaries of Dade County as well as the regional boundaries incorporating Dade, Broward, and Monroe Counties, and the watershed boundaries used by the South Florida Water Management District, another regional entity that is important in South Florida. This chapter begins by describing the region in which Miami is situated, both in Dade County and in the larger regional context of South Florida, which is increasingly economically, environmentally, and legally intertwined. The metropolitan government of Dade County then is discussed, followed by descriptions of other regional entities and their importance to the area.

■ Miami/Dade County in the 1990s

Miami is the largest city in Dade County as well as in the entire South Florida Region, containing 10.7% of the region's population (359,973 people) within its borders in 1992. In comparison, Fort Lauderdale, the county seat and the largest city in Broward County, is only 40% the size of Miami, with a population of 147,678. For purposes of this chapter,

Dade, Broward, and Monroe Counties are considered to be South Florida (the state of Florida definition). In some informal usage, Palm Beach County would also be included in South Florida (South Florida Regional Planning Council, 1993b).

In 1992, Dade County (with 1,982,901 people) contained 25% of the population of the state of Florida (Figure 8.2). Of the three counties officially composing the South Florida region, Dade County is the largest, followed by Broward County, then Monroe County, which is the Florida Keys. The unincorporated portion of Dade County, the largest subcounty entity within the region at 1,060,739 people, is practically three times the size of Miami and third behind Dade County itself and all of Broward County in population among the region.

Two demographic trends that have shaped the Miami/Dade County area since the 1950s are the growth experienced in the area, with its accompanying environmental and growth stresses, and the arrival of the Cuban American community after the Cuban Revolution in 1959. The population growth in Miami, Dade County, and South Florida over the past 40 years has created the context for much of the regional politics in the area. The region exhibited enormous growth in the 1950s, as Miami became urbanized with the return of veterans stationed there during World War II. This growth was mirrored in South Florida as well as the entire state of Florida, among the fastest growing areas in the country at that time. Since then, the growth has continued, although Dade has grown more slowly than either South Florida or the state of Florida.

Vast economic and cultural changes in Miami and Dade County can be seen in the racial and ethnic differences occurring in the region over time. Although large immigrant influxes to the area have had detrimental effects for the area, the arrival of the Cuban community also has had undeniable benefits for South Florida, helping to transform Miami and South Florida into the gateway to Latin and Central America. Following the 1959 revolution in Cuba, a large number of middle- and upper-class Cubans left their home country for exile; many came to Miami via air or boat. Once there, they formed an increasingly influential economic and political bloc in the area.

Between 1980 and 1990, Miami grew from 55.97% to 62.5% Hispanic, while the percentage of Hispanics in Dade County increased from 35.7% to 49.2%. Many Hispanics moved to unincorporated Dade County, as the percentage there grew from 26.2% to 41.5% in a decade. By 1990, Miami contained 20% more Hispanics than the state of Florida and over 10%

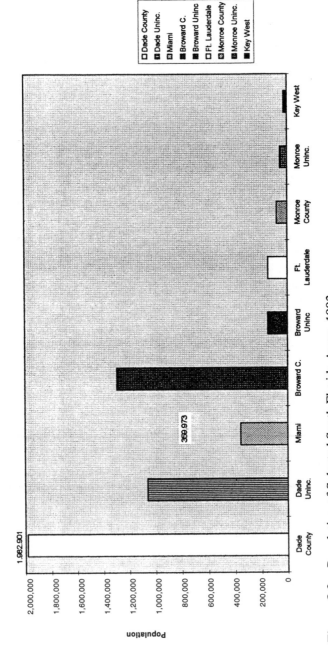

Figure 8.2. Population of Selected South Florida Areas, 1992

SOURCE: *South Florida Profile*, 1993.

more than Dade County (Research Division, Metro-Dade County Planning Department, 1993).

Dade County as a whole is a much more affluent area than Miami itself. In 1990, Dade County had a median family income of $31,113, whereas Miami's median family income was only $19,725. Thirty-one percent of Miami's residents lived below the poverty line, whereas only 18% of Dade County, 14.8% of South Florida, and 12.7% of the state of Florida lived below the poverty line (Research Division, Metro-Dade County Planning Department, 1993).

One of the benefits of the Cuban and other South and Central American migration into Miami in the last decades, along with the efforts of Miami's leadership, has been the development of Miami into the "gateway to Latin America." As such, there is an increasingly international bent to the region's dominant economic sectors. According to the South Florida Regional Planning Council,

> More than 300 multinational companies have opened world, regional and Latin American sales, service, administrative, research, training and manufacturing facilities in Dade County. In 1991 alone, $21.7 billion worth of goods—$13.4 billion in exports and $8.3 billion in imports—was processed through the Miami Customs District, which include airports and seaports in Miami, Fort Lauderdale, West Palm Beach, Fort Pierce and Key West. This represents an increase of 110 percent over the totals for 1985. Over 70 percent of that trade is with Latin America and the Caribbean. . . . Tourism contributed $7.2 billion to the Miami/Dade County economy in 1991. . . . About 31 percent of the area's workforce directly or indirectly benefits from the visitor industry. . . . Miami/Dade County has 46 foreign banks agencies and 15 Edge Act banks, representing 36 percent of the local banking institutions. Miami/Dade County has the highest concentration of foreign bank agencies in the southeastern U.S. (South Florida Regional Planning Council, 1993a, p. 4)

Dade County has suffered several extreme stresses to its local economy and polity in recent years. First was the 1980 influx of more than 125,000 Cuban refugees, an event that threatened to reoccur in August, 1994. The acceptance of that many refugees over a short period of time generated severe strain on Dade County's ability to provide services for its citizens, as well as the newcomers. The second, also severe, blow to this international and tourism-based economy occurred on August 24, 1992, with the landfall of Hurricane Andrew. South Dade County suffered the most from the hurricane, as many businesses were at least temporarily closed and many employees out of work. Tourism was damaged, and Homestead Air Force Base, with a large civilian workforce, was temporarily closed. Another of the area's critical industries, agriculture, suffered $1 billion in damage, including $250 million in permanent income loss and $580 million damage in lost structures (South Florida Regional Planning Council, 1993c).

■ Metropolitan Dade County: Fulfillment of a Promise?

Since 1957, Dade County has had a modified two-tier system of government called Metropolitan Dade County, or simply "Metro." The system is a modified one because the citizens in the unincorporated portions of Dade County receive all of their services from Metro (the upper tier) and so have only one tier of government at the local level, whereas citizens in the 27 incorporated cities remaining in the county receive services from both their city (the lower tier) and from Metro.

The South Florida region is a relatively young urban area. The first Dade County Commission was not elected until 1877, and the City of Miami was not incorporated until 1896; Miami Beach was not incorporated until 1915. A series of land booms and busts, particularly during the 1930s, was responsible for much of the population growth in Dade County. This growth continued during the 1950s with returning World War II veterans and the Sunbelt boom of the 1960s and 1970s. Particularly during the 1950s, population growth was enhanced by the drainage of the Everglades for living space and with the development of a year-round tourist season (Lotz, 1984). In Dade County, the Sunbelt growth of the 1960s and 1970s was accentuated by the large number of exiles arriving from Cuba; this pattern of immigration was repeated with migrants from other Latin American and Caribbean nations arriving during the 1980s and 1990s (Lotz, 1984).

Even during the 1940s, some stresses in local governments' ability to deal with current problems began to show, and leaders began consolidation of some services, such as public health, education, tax assessment and collection, and airport and seaport development. These were considered appropriate because most of the county's municipalities were situated directly around the borders of Miami, which even then was the largest city in the county (Carter, 1974).

Even these changes did not relieve the stress on local governments. During the 1950s, more and more of the population moved into the unincorporated areas. This growth created all the problems of an urbanized area for a county government ill equipped to provide services for such an urban population. City elites became concerned that their governmental structures would not be able adequately to manage the growth that was expected to follow, particularly outside the central city of Miami (confidential interview, 1992).

A series of measures to reform local government followed, including a 1953 measure that Miami voters barely defeated that would have consolidated and abolished the City of Miami and transferred all functions to the county (confidential interview, 1992; Carter, 1974). These sentiments continued, as many citizens at the time wanted to fully consolidate the county and do away with all cities. Local analysts suggested that the "county government was created in territorial days . . . its form has changed very little since the adoption of the State Constitution in 1885" (Lotz, 1984, p. 27) and "the multiplicity of municipalities and their illogical and frequently weird arrangements is a serious impediment to the administration of local government" (Hertz, 1984, p. 8).

A local government study in 1954 found widely varying levels of services and taxes across the cities in Dade County and poorly managed services from the county itself. Recommendations in the study included creating a governmental structure capable of handling areawide problems and dealing with the growing fragmentation of governmental bodies in the county (Lotz, 1984). The study report went on to describe the rationale for the structure that eventually became Metropolitan Dade County, the creation of a federation, an areawide government for the entire county with cities maintaining authority for certain services. Under this federated system, cities continued to exist, but an areawide government also was provided.

So began the controversy, still continuing, over the structure of Dade County and existing municipalities. The Dade County Home Rule Amendment to the State Constitution was passed by Florida voters in 1956, with 71% of the voters in Dade County (Carter, 1974) approving the measure. Following two Charter Boards and public hearings, and backed by a coalition of the Miami-Dade County Chamber of Commerce, the League of Women Voters, the *Miami Herald,* the *Miami News,* and the Dade delegation to the state legislature, and amid much controversy and discussion, the voters of Dade County adopted the final charter in 1957 (Carter,

1974). With a very small turnout, 51% of the voters casting ballots (26% of the eligible voters in Dade County) approved the creation of Metropolitan Dade County (Lotz, 1984).

The Two-Tiered Structure

The charter left the county and 26 incorporated cities in existence (one more, Islandia, was incorporated by the Metro Commission in 1961). The county already had a consolidated school system, which has remained a separate entity, and the court system also remained unchanged. By vote, the outmoded Dade County form of government had been changed into a system that provided county-level services and regulatory action for all citizens within the county (the upper tier) and municipal services for those citizens living in the unincorporated areas while retaining some municipal level (lower tier) services.

The charter established an 11-member County Commission, elected from districts (since changed to 9 commissioners elected at large but required to live in districts, then back to 13 districts under court order in 1993), with policymaking and legislative—rather than their earlier administrative—powers. Like earlier urban reforms, this switch encompassed the creation of the position of county manager; the county manager performed duties such as county reorganization and budget preparation.

Table 8.1 illustrates Metropolitan Dade County's two-tiered approach to service delivery. On the top tier are services provided by the county. These services include traditional county-level services such as tax collection and appraisal, election administration, the provision of regional parks and recreation, and public health or metropolitan services. Metropolitan services, also part of the top tier, are those such as public transportation and traffic engineering provided areawide by Metropolitan Dade County in order to take advantage of the economies of scale of this large area.

County services, also part of the top tier, were largely those given to the county to prevent duplication. The only charter-mandated merger of services that remained was property assessment and tax collection. To prevent duplication, the charter provided that the county tax roll would be the only allowable roll and that no municipality would be allowed to do its own assessments. The County Commission also utilized its authority to abolish the City of Miami Housing Authority in 1968 so that now, the county provides housing authority services to all but three cities.

TABLE 8.1 Service Provision in Dade County

County Services—Provided Countywide by Metro

- Law Enforcement, Courts, and Medical Examiner
- Tax Collection and Appraisals
- Elections
- Some Public Works
- Regional Parks/Recreation
- Civil Defense
- Emergency Assistance
- Public Health
- Agriculture Services

Metropolitan Services Provided by Metro

- Public Transportation
- Public Works
- Environmental Protection, Water/Sewer, and Animal Services
- Traffic Engineering
- Solid Waste Disposal
- Libraries
- Fire/Emergency Rescue Services
- Development Planning/Housing
- Regulatory Services
- Airports, Seaports
- Parks/Recreation and Cultural Services
- Mental Health Services

Comprehensive Services Provided by Metro to Unincorporated Areas

- Garbage and Trash Collection
- Zoning Enforcement
- Neighborhood Parks and Recreation

Services Retained and Provided by Cities

- Police Patrol/Some Law Enforcement
- Zoning Enforcement and Planning/Zoning
- Neighborhood Parks and Recreation
- Fire Protection (4 cities)
- Libraries (9 cities)
- Housing Authorities (3 cities)

Functions such as election administration, civil defense, public health, and emergency assistance also are services that should not be duplicated and as such were moved to Metro Dade.

Metropolitan services are those areawide services that Metropolitan Dade County gained under the new charter and performed for all of Dade County. These services differed from the county services in that they were not traditionally performed by a county government and were not provided by a series of special districts, as they are in many communities. The services include public transportation, environmental protection, water/sewer, traffic engineering, regulatory services, airports and seaports, development planning and housing/community development, and solid waste disposal. Traffic engineering became the first of the "Metro" services provided by the new county organization, as the few cities that had capacities in this area quickly transferred them. In addition, the only cities to have incinerators for solid waste disposal, Miami and Coral Gables, soon turned them and the responsibilities for waste disposal over to Metro.

Regional transportation is a good example of a service adopted by Metro because of its regional, areawide implications. Under Metro, a unified bus system operates throughout the county, as does MetroRail, a fast-rail system. In addition, the public works functions of arterial streets and traffic engineering were given to Metro so that a unified set of standards would be utilized throughout the county and so that areawide planning and engineering could more easily take place. Likewise, housing and community development were determined to be metropolitanwide functions and so were placed under Metro rather than retained by individual cities.

Some regulatory services are provided by Metropolitan Dade County to ensure uniform application. Although cities may establish their own regulations in various arenas, these may not violate those of Metropolitan Dade County; they may only exceed Metro's own standards. To ensure uniformity, one of the earliest actions by the Metro government was to pass a series of ordinances applying to the entire county. These included a uniform traffic code superseding all city traffic codes, examinations for construction personnel, a uniform subdivision ordinance and the South Florida Building Code, a criminal code, and a dog control ordinance. Challenged in court, these early ordinances provided the opportunity to test the legality of the powers of the new government (Lotz, 1984; Carter, 1974).

The lower tier of services includes the services still provided by cities and those comprehensive services provided by Metropolitan Dade County to the unincorporated areas. Services provided by cities include municipal zoning, police patrol, some other law enforcement services, and neighborhood parks and recreation. Some cities still provide their own fire protection, libraries, and housing authorities. Libraries are one of the services selectively provided by cities, with nine cities choosing to retain some library services but 80% of the county served by the Metro-Dade Library System. Fire protection and emergency medical rescue is another service selectively provided. Four cities—Miami, Miami Beach, Hialeah, and Coral Gables—maintain their own fire protection services, but Metro now serves approximately 60% of all county residents (Lotz, 1984).

Metro provides some of these same services to the unincorporated areas, including neighborhood parks and recreation, zoning enforcement, and garbage and trash collection. It is important to note that the Metro Dade Commission acts as the city commission for the unincorporated areas. Originally, Metro took responsibility for zoning and building regulation as well as public works; however, the outcry from the cities was so strong that Metro commissioners repealed those ordinances in 1958 and gave zoning code development back to the cities (Carter, 1974).

■ Metropolitan Dade County—A Success?

Since the beginning of Metro, there has been controversy over the new governmental structure and its effectiveness. This controversy was enhanced by its seeming complexity and the lack of a clear mandate for its initial creation. Even with such unclear beginnings, however, Metro began its life with a positive performance rating of more than 80%. The level of public support for Metro has wavered over the years, however; this high level of approval declined steadily to 36% in 1963, then increased again to 64% in 1973, the date of the last comprehensive set of public approval studies (Hertz, 1984).

Citizen Satisfaction

The relationship between Metro and the cities, particularly Miami, always has been a conflictual one. This is reflected in the fact that the percentage of voters who thought the relationship between Metro and other governments was "about right" only went above 50% one time for any group during the period from 1962 until 1973. By geographic area, 26% of all county voters thought the relationship was about right in 1962; this peaked at 43% in 1968. Significantly, 53% of Miami voters thought the relationship between the cities and Metro was "about right" in 1968, whereas only 36% of unincorporated voters agreed (Hertz, 1984). Frequently, differences emerged in the views on Metro between Miami residents and residents of other Dade County municipalities. Although the differences fluctuated over time, whenever there were differences, voters were less satisfied with Metro than with other governments in the county (Hertz, 1984).

By 1987, other studies found significant differences by ethnic group (Brierly & Moon, 1988a). Only 27% of non-Latin whites thought Metro performance in 1987 was good or very good, whereas 43% of Hispanics and 41% of Blacks approved. These differences are especially significant and surprising given the significant degree to which Hispanics and African Americans historically had been underrepresented on the Metro-Dade Commission. Given these electoral differences, it is also surprising to see that a large proportion of Hispanics thought they had a very good or good "voice" in Metro government (39.8%), compared to 30.3% of non-Latin whites and 36.2% of African Americans. The perceptions of treatment by Metro, however, certainly follow expectations for African Americans, given their electoral and appointed representation patterns. Fifty percent of non-Latin whites thought that treatment by Metro was very good or good, 52.1% of Hispanics felt treatment was good or very good, but only 25.3% of African Americans felt their treatment by Metro was good or very good (Brierly & Moon, 1988b).

Is It Representative?

These perceptions reflect the fact that, over time, Metro Dade County has not been as reflective of its diverse community as some Dade County municipalities such as Miami or Hialeah. In a community like Miami and Dade County, where ethnic and racial divisions are highly salient and color many governmental transactions, Metro-Dade's relative lack of diversity and responsiveness was problematic. As seen above, one reason for the divergent views of ethnic groups about Metro-Dade County government has been the perception and evidence of bias. Until a Voting Rights Act lawsuit brought by African American and Cuban elected officials and candidate plaintiffs in 1986 (*Carrie Meek, et al., v. Metro-*

politan Dade County, Florida, et al.) was settled by the county in 1993, only one minority group member had ever been elected to a seat on the commission without first being appointed to a vacancy (Citizens Charter Review Committee, 1986). At the time of the 1986 lawsuit, only one African American (Barbara Carey) and one Cuban American (Jorge Valdes) held seats on the nine-member County Commission. Both of these members initially had been appointed to their seats and were subsequently reelected.

With the settlement of the Voting Rights Act case in 1993, the Commission was increased to 13 seats and elections were moved to a single-member districts system, a reform long discussed and put before the voters numerous times. In the first elections with this system, four African Americans (including Arthur Teele, Jr., the head of the federal Mass Transit Agency in the Reagan Administration) and six Hispanic Americans were elected to seats on the Commission.

Not surprisingly, with this history of a lack of electoral representation, affirmative action statistics indicate that not only electoral representation of ethnic and racial groups was lacking in Dade County. The hiring of African Americans and Cuban Americans in the Metro-Dade bureaucracy also lagged behind the proportions of ethnic and racial groups in the population at large, as well as those of the City of Miami, with the absence of pressure from Hispanic or African American elected officials. Affirmative action statistics indicate that the proportion of Hispanics hired by the City of Miami (with three of five Hispanic City Commissioners) was practically 10% higher than the proportion of Hispanics hired by Dade County (with only 11% [one of nine] Hispanic Commissioners). Clearly, without the benefit of the Voting Rights Act lawsuit, Metro-Dade County was not as responsive to the ethnic diversity of the majority of its citizens as was the City of Miami.

Unincorporated Areas

Beyond issues of representation lies the question of the effectiveness of Metro Dade County as a regional or even countywide governing body. Only 35% of the county's population resided in the unincorporated area when the charter was adopted (Citizens Charter Review Committee, 1986). By 1970, 42.4% of the county's population lived in the unincorporated area. The unincorporated area within Dade County has grown steadily since that time and promises to continue growing (up to 60% of the population will live there by 2005, according to the Metro Dade

Planning Department). By the 1990 census, more than half (53.1%) of the county's population lived in the unincorporated areas and so was governed solely by Metro Dade County. This demographic shift puts the entire notion of a two-tiered government into question, as decreasing numbers of Dade County's citizens are living under the federated system intended for the entire county.

Having more and more of the county population under the governance of Metro Dade County puts a strain on the resources of the Metro Dade County Committee and staff that was never intended by early advocates; they designed the system so that representation and some services would be provided by entities besides Metro Dade. As the Citizens Charter Review Commission stated in its final recommendations in 1984, the County Commission spent more than half of its time on zoning issues for the unincorporated areas, including "considerable time debating issues such as whether a particular property owner should be granted a variance to construct lights on his backyard tennis court" (Citizens Charter Review Committee, 1986, p. 14).

The representation issue is also connected to the issue of zoning at the local level, one of the most controversial services still provided by the municipalities. At the beginning of the Metro Dade experiment, Metro had control over local zoning; however, there was such an outcry from municipalities that the Metro Commission gave the authority to conduct zoning back to local governments. This situation has changed, however, as growth in the county has been in the unincorporated areas rather than the landlocked and sometimes built-out municipalities. In fact, Dade County is cited as one of only a few counties utilizing a countywide planning process in a state oriented and mandated toward regional planning (Paterson, 1988). Some citizens and analysts believe that a regional or countywide government is too removed to be able to be influenced or to allow citizens or builders to be heard and that the County Commission is too overburdened to deal with the details of zoning for such a large and growing area.

As the system increases noticeably in complexity and fragmentation, the very motivation behind the federated Metro system during the 1950s is beginning to dissipate. One solution to this problem is the incorporation of additional cities in the increasingly urbanized unincorporated areas. This would allow some citizens to have a government that is closer to them as well as decreasing the size of the governing task for Metro Dade. The Metro Dade County Commission has actively resisted incorporating more cities within the county (Citizens Charter Review Committee,

1986). Only one small city, Islandia (part of the Florida Keys), has been incorporated since Metro-Dade's inception.

Yet another concern is the development of future political leadership. With only county commission seats and 27 municipal councils to run for, some people are concerned that political interest and opportunity for the average citizen would fade. Thus, an important source of new political leadership would be reduced.

Metro Dade County thus has become more of a regional government covering a broad area, rather than simply a county or even a metropolitan county government, whether or not it is capable of providing effective regional and comprehensive government to all of Dade County. It is less and less the type of government that Dade reformers and voters originally intended.

Structural Issues

Along with single-member districts, two other Metro-Dade County structural issues, partisan elections and a strong mayor, have persisted and dominate any discussion on the effectiveness of Metro Dade County government. The issue of single-member districts and full representation for Metro Dade voters has now been decided by the Voting Rights Act lawsuit.

Proponents of single-member districts claimed that, with only two members of minority groups represented on the County Commission, the only way to ensure adequate representation was to have single-member districts. Opponents of the plan countered that under the 1991 plan "Dade voters would get to elect only the mayor and two of the 13 commissioners —one from a district, the other from a region. Now, voters elect all eight commissioners and the mayor" (Epstein, 1991, p. 1A). Some African American activists believed "this plan is an emasculation of any kind of black vote. . . . You do not have a full vote; you have three-thirteenths of a vote" (Epstein, 1991, p. 12A). Still others thought that districts would increase the contentiousness of the already conflict-ridden ethnic rivalries in the county and would lead to parochialism and ward politics, the classic arguments against single-member districts. With 10 of the 13 new Commissioners either Hispanic or African American, in the short term it appears that single-member districts have been effective. It remains to be seen whether these newly elected representatives will represent their respective groups effectively and whether structural and policy changes will be made.

The issue of a strong mayor is also one that has remained over the years and has been on the ballot several times. This issue, most salient between 1982 and 1992 (Fiedler, 1985b), led to the establishment of the Citizens Charter Review Committee, the group of political and economic elites who developed recommendations in an attempt to make the changes in the Metro Dade Charter they thought were needed. This committee, funded by several large corporations in the community (Ryder System Inc., the rental trucking group; Knight-Ridder Newspapers Inc., publisher of the *Miami Herald;* Southern Bell; and Florida Power and Light Co.), became seriously concerned with the structure of Metro Dade when the crime issue arose. In particular, business leaders became more concerned with the strong mayor system when they perceived that Mayor Steve Clark did nothing when snipers fired on motorists during the summer of 1985 (Dugger, 1985).

Other activists, such as Annie Ackerman, leader of the North Dade County condominium associations, a very powerful electoral voting bloc, believed that "Citizens would have more influence with an elected official than a manager. . . . If they don't like the mayor, they can vote him out of office" (Dugger, 1985, p. 5B). Suggesting the elite nature of this issue, yet another member of the Citizens Committee thought "there was no huge boiling pot of discontent about county administration. I don't detect that people are ready to storm the brand-new administration building" (Dugger, 1985, p. 5B).

The issue is moot for the time being, as Steve Clark, who was criticized consistently for his weak leadership as mayor of Metro Dade, ran and won election as mayor of Miami, where he previously had served on the City Commission. Under the new system, the Board of County Commissioners has no mayor, even a weak one, but only a chair and vice chair. The new chair is Arthur Teele, Jr., an African American; the vice chair is the former mayor of the City of Miami, Maurice Ferre, who is Puerto Rican. Ferre distinguished himself as a strong but controversial leader while mayor of Miami, which also has a structurally weak mayoral system. Both of these leaders are experienced in other political arenas and will have the opportunity to show that a weak mayoral or commission system may still produce strong leaders.

Has It Worked?

A longtime government activist and analyst stated, when asked whether Metro Dade can be considered a "true" metropolitan government, certainly at the time that it was created and for many years afterward, it is a metropolitan government. It does govern the entire county regardless of the existence of municipalities. The major legislation, with the exception of zoning legislation, is countywide in its effect. There is a strong county-wide planning mechanism. It is flawed. A major flaw, in my opinion, is that there is no secondary level of government in the unincorporated area. [But] it would have been political chaos if it hadn't been established. Definitely, I think it has worked. It has not worked in its two-tier aspect in the unincorporated area, and I think this is becoming better recognized now, particularly in the zoning end of things. (confidential interview, 1992)

With two contentious structural issues finally resolved after years of controversy, and with strong, representative leadership on the Commission, Metro may yet become an effective regional government. Metro Dade really can no longer be considered a two-tier system, as originally intended. Instead, it is something between a strong county or metropolitan county government and a truly regional government.

■ The "Other Region"

Although Metro-Dade County has accomplished some regional governance for Dade County, Miami and Dade County are also part of a broader region, South Florida. As Carolyn Dekle, executive director of the South Florida Regional Planning Council, suggests, the region is beginning to "clearly articulate and buy into a regional identity . . . we think of ourselves locally and do everything regionally" (personal interview, 1993). The move toward regional cooperation can be seen particularly in the efforts to manage the aftermath of Hurricane Andrew, which affected the entire South Florida region, and a new project proposed for the borders of Dade and Broward Counties. This project, proposed by Wayne Huizenga of Blockbuster video store fame, would be a huge sports complex close to the site of Joe Robbie Stadium, home to the Miami Dolphins and the Florida Marlins. One example of the attempt to move toward regional problem solving is the creation of a private regional entity, the Tri-County Athletic Association, to work through some of the many issues created by this proposed project.

Because of progressive Florida state environmental and growth management legislation passed since the 1970s, local governments such as those of Miami and Dade County are required to respond to these larger regions through quasi-governmental bodies, regional planning councils and water management districts, outside their own borders. This creates policy imperatives and politics that go beyond Dade County.

Although not adequately funded, areas of critical state concern (ACSC) and developments of regional impact (DRIs) are two tools designed to protect Florida's natural environment. In effect, they created a framework for addressing development, environmental, and growth issues at a level greater than that of counties. Areas of critical state concern are geographic areas with environmental characteristics that require intervention by the state. Related to the ACSC are Resource Planning and Management Committees designated for areas that are endangered environmentally in some way. South Florida contains one ACSC, the Florida Keys, and Dade County has one RPMC, the Everglades. These designations, made by the governor and cabinet, allow resource plans to be developed that are themselves then passed by the governor and cabinet. Once the plans are adopted, government agencies at all levels are then expected to work to implement the plan elements, mandating that the state or regional perspective be followed by municipal and county governments. This policy initiative inserts a regional perspective into local land use. For example, after the Everglades management plan passed in 1985, local governments, including Dade County, were expected to make changes in their local plans and plan implementation to be consistent with the management plan (DeGrove & deHaven-Smith, 1988).

Another regional policy entity, developments of regional impact (DRIs), are any development that would have a substantial effect on the health, safety, or welfare of the citizens of more than one county through their size, nature, or location (O'Connell, 1986). The effects of such developments would be reviewed and evaluated for their regional impact by one of the 11 regional planning councils, established in 1973. This process alone gave the regional planning councils a significant regional role and strongly established a regional presence in the local development process (Starnes, 1993).

Subsequent state legislation has attempted to strengthen the role of regional planning and regional planning councils. Local governments are now mandated to develop local plans, which are then reviewed by the regional planning councils, although their role is advisory and few resources are provided to implement the policy. Today, regional comprehensive plans are required by the state legislature and are adopted by administrative rule after being reviewed for consistency with the state comprehensive plan, also adopted by the state legislature. In this top-down process, these plans now provide the basis for review of DRI applications as well as for all other agency actions (Starnes, 1993). The role of the regional planning councils in reviewing local plans remains advisory—the RPCs review local plans for compliance.

The capstone of years of effort to insert regional and state perspectives into local government and development was the 1985 Omnibus Growth Management Act, which required consistency, concurrence, and compactness among local and regional plans. The consistency requirement, the most important of these provisions, established an integrated policy framework for the state, beginning with the goals and policies of the State Plan. The plans of state agencies, then regional and local agencies, were all required to be consistent with the State Plan and with the plans of the agency above them. This called for a shift in perspective for local government. Not only is local planning without a broader perspective not allowable, but essentially local powers have been ceded to the regional government.

Yet another regional element are the five Water Management Districts (WMD) in Florida. These bodies have borders concurrent with the watersheds of major rivers; they have limited taxing authority and regulate water use and surface water management, with authority over wetlands and water protection issues. They were established for flood protection purposes. A full canal system was developed by 1967, and the South Florida Water Management District alone contains 700 canals over 16 counties, including Dade County. According to Thomas Singleton of the South Florida Water Management District, for the past 5 to 10 years, the district has been attempting to move beyond a past role perceived as too sensitive to agricultural needs to one that balances all needs and interests into an integrated role of water supply planner (personal interview, 1993). Water management districts can be very powerful players in the arena of local environmental and water supply planning and implementation. These entities and the water supply planning role, according to Singleton, place the regional perspective squarely over the local perspective for Miami, Dade County, and other local governments.

■ Conclusion

As in many other areas in the United States, more and more elements of regional problem solving and governance are emerging in Miami and Dade County today. Metro Dade County, along with the regional planning bodies in the state of Florida, regional planning councils and water management districts, has changed the way traditional local governments and citizens go about their daily lives. Metro Dade County itself has become increasingly regional in its governance as more of Dade County's population moves into the unincorporated areas, accompanied by the Metro Commission's reluctance to incorporate additional cities. The region Metro Dade governs is still merely a county, not the full region. Although many consider Metro Dade County a regional government, it is more accurately described as neither a regional government nor as merely a metropolitan county government. Instead, it is a hybrid, perhaps called a "regional county" government. Its role goes beyond that of a traditional county government and even beyond a metropolitan county government in that it provides the entire county with some services and also supplies metropolitan services often provided by special districts. It is not quite a regional government because its powers stop at its limited borders. It provides some regional services but only within one county; therefore, it is more of a "regional county" structure.

Many citizens have doubted the capacity of Metro, even as a county-level government, to address their concerns as well as a purely municipal government. Recent events suggest a more optimistic appraisal of Metro's potential. Ongoing disputes—district elections and a strong mayor—have now been resolved in favor of single-member district elections and a Commission chair/vice chair system. A majority of minority group members have been elected to the Commission under the new system, increasing the system's legitimacy and credibility. The perceived leadership capacity of Metro Dade government already has improved with the selection of strong leaders as chair and vice chair. Given the growing emphasis toward regional problem solving in the area and these positive changes within Metro, Metro finally might have the support and the capacity to provide strong, effective government in its diverse and complex community. It remains to be seen whether that will actually happen.

REFERENCES

Brierly, A. B., & Moon, D. (1988a). *The effect of ethnic cleavages on metropolitan reform in Greater Miami* (Working Paper). Coral Gables, FL: University of Miami School of Business Administration.
Brierly, A. B., & Moon, D. (1988b). *Hispanic attitudes toward metropolitan reform in Greater Miami* (Working Paper). Coral Gables, FL: University of Miami School of Business Administration.

Carter, L. J. (1974). *The Florida experience: Land and water policy in a growth state.* Baltimore: Johns Hopkins University Press.

Citizens Charter Review Committee on the Dade County Charter. (1986). *The final recommendations of the Citizens Charter Review Committee on the Dade County Charter.* Unpublished manuscript.

DeGrove, J. M., & deHaven-Smith, W. J. (1988). Resource planning and management committees: Implementing Florida's Critical Area Program. In W. J. deHaven-Smith (Ed.), *Growth management innovations in Florida* (pp. 135-152). Fort Lauderdale, FL: FAU/FIU Joint Center for Environmental and Urban Problems.

Dugger, C. W. (1985, November 7). Civic heads seek to boost metro mayor's powers. *Miami Herald,* p. 1B.

Dugger, C. W. (1986, October 5). Ethnic politics comes of age in Dade. *Miami Herald,* pp. 1B, 4B.

Epstein, G. (1991). At stake: Blueprint for Metro. *Miami Herald,* pp. 1B, 12B.

Fiedler, T. (1985a, May 23). Parties: Make races for metro partisan. *Miami Herald,* p. 2B.

Fiedler, T. (1985b, November 7). Strong mayor plan: A rocky road. *Miami Herald,* p. 2B.

Hertz, D. B. (1984). *Governing Dade County: A study of alternative structures.* Unpublished manuscript, University of Miami.

Lotz, A. (1984). *Metropolitan Dade County: Two-tier government in action.* Boston: Allyn & Bacon.

O'Connell, D. W. (1986). New directions in state legislation: The Florida Growth Management Act and state comprehensive plan. In J. M. DeGrove & J. C. Juergensmeyer (Eds.), *Perspectives on Florida's Growth Management Act of 1985* (pp. 48-63). Washington, DC: Lincoln Institute of Land Policy.

Paterson, R. (1988). Coordinating local government comprehensive planning: A mandate and necessity. *Florida Environmental and Urban Issues, 16,* 8-16.

Research Division, Metro-Dade County Planning Department. (1993). *Dade County facts.* Unpublished manuscript.

Sofen, E. (1963). *The Miami metropolitan experiment.* Bloomington: Indiana University Press.

South Florida Regional Planning Council. (1993a). *Our regional identity: Meeting the challenge.* Hollywood, FL: Author.

South Florida Regional Planning Council. (1993b). *South Florida: A statistical profile.* Hollywood, FL: Author.

South Florida Regional Planning Council. (1993c). *South Florida: Global community—background report.* Hollywood, FL: Author.

Starnes, E. M. (1993). Substate frameworks for growth management. In J. M. Stein (Ed.), *Growth management: The planning challenge of the 1990's* (pp. 76-95). Newbury Park, CA: Sage.

9 Minneapolis-St. Paul: Structuring Metropolitan Government

JOHN J. HARRIGAN

If imitation truly is the sincerest form of flattery, residents of the Minneapolis-St. Paul metropolitan area have reason to wonder about the praise that has been given to their system of metropolitan governance over the years. "Widely praised but never copied" easily could be the logo put above the door of the Twin Cities Metropolitan Council. For the past 15 years, the Metropolitan Council probably has received more praise from outside the Twin Cities than from within it.

However prominent the Metropolitan Council might look from afar and however successful it was in the past, there is much uneasiness in the Twin Cities today about whether the Metropolitan Council can provide direction on three overriding challenges—the need to position the region to compete effectively in the changing global economy, the need to alleviate the growing central city/suburban disparities that are seen to threaten the region's quality of life, and the need to gain better control over suburban sprawl.

This chapter will address these concerns by sketching out the challenge of global competition, outlining the patterns of central city/suburban disparities and relating these to the Twin Cities governance model, explaining a growing disenchantment with the governing model in the 1980s, and describing significant changes that were made in the model in the 1990s. In making these assessments, this chapter will distinguish between a narrow and a broad definition of the region. Narrowly, the term "metropolitan area" will be used to mean the seven counties that fall within the jurisdiction of the Twin Cities Metropolitan Council. Broadly, the term "region" will be used to mean the much broader land mass that falls within the economic and cultural sway of the Twin Cities. This broader region covers the entire state of Minnesota and nearby portions of four neighboring states.

■ Positioning for Global Competition

A useful guidepost for assessing an area's comparative advantage is John Mollenkopf's (1983, pp. 31-36) identification of three different adaptations to the nation's transformation from an industrial to a post-industrial economy. Some old industrial cities, like Detroit, failed to make the transformation and have gone into severe economic decline. Newer cities, such as San Diego and Phoenix, had the good fortune to hit their growth phase precisely when the postindustrial economy went into high gear, and thus they virtually grew up as service and/or administrative sites for high-technology industries. Still others, such as Boston and San Francisco, transformed themselves from dependence on industry into significant service centers for finance, medical care, and corporate services.

The Twin Cities area is a moderately successful example of the third type of adaptation. During the early 20th century, its economy was based heavily on manufacturing, railroading, grain milling, and the stockyards. All these sectors declined over the past four decades, but the region itself prospered over this time frame. There currently is great concern over the ability of the metropolitan area and the broader economic region to continue adapting successfully to the changing global economy, and important local organizations have sponsored studies that are addressing this issue from various angles.

Positive Signs for Global Positioning

The Twin Cities possess several economic advantages. Minnesota has enjoyed above average job growth in the North Central region (U.S. Bureau of the Census, 1993, p. 418) and below average unemployment rates (*New York Times,* June 30, 1994, p. C2). The state was ranked by the Corporation for Enterprise Development as one of only eight states best poised for future economic growth (*Minneapolis Star Tribune,* June 30, 1994, p. C2). The state has also been a dynamic leader in medical services innovation. Among the medical technology firms located in the Twin Cities, for example, are the nation's premier manufacturers of artificial

heart valves and heart pacemakers. The region houses two leading research medical complexes at the University of Minnesota Hospitals and
the nearby Mayo Clinic. Minnesota's overall economic output is fairly
consistent with its size. Ranking 20th in population, the state ranks 19th
in gross state product and 17th in median household income (U.S. Bureau
of the Census, 1993, pp. xii, xix, 444).

Despite an outmigration of retirees to the Sunbelt climes of Florida and
Arizona, the Twin Cities have experienced a net in-migration of people.
Most of these come from the five-state region of Minnesota and its
immediate neighbors (Sternberg, 1993). The vast majority of counties in
the five-state region lost population in the 1980s, and a large portion of
the rural émigrés from these counties wound up in the Twin Cities suburbs
(*Minneapolis Star Tribune,* March 9, 1991, p. 5B).

Finally, the Twin Cities metropolitan area dominates this broader
region economically and culturally. It is the regional mecca for financial
services, advanced medical services, long-range air travel, upscale shopping, and entertainment. Major league sports are the most visible symbol
of this dominance. There is no major league sports team outside the
metropolitan area within a radius of more than 300 miles. Because of its
geographical location, Minneapolis-St. Paul is not overshadowed by giant
neighbors the way that Milwaukee is overshadowed by Chicago or San
Jose is by San Francisco. Furthermore, there are hopes that the 1993 North
American Free Trade Agreement will dramatically stimulate north-south
trade routes as distinct from the historic traditional patterns of east-west
trade routes. If this happens, it could give an important economic boost
to midcontinent metropolises such as Minneapolis-St. Paul, which serve
as gateways to the north-south trade routes.

Much of this positive picture is summarized in Table 9.1. Among the
25 largest MSAs, both the central cities and the suburbs of the Twin Cities
compare favorably in terms of education, income, poverty rates, and
violent crime rates.

Ominous Signs for Global Positioning

In contrast to these positive signs of regional dynamism, there are also
some ominous signs. The geographic position that enables the Twin Cities
to dominate a large region also isolates manufacturing businesses from
large consumer markets. In comparison to other states, Minnesota is only
average in its share of U.S. exports. (U.S. Bureau of the Census, 1994,
p. xxii). Minnesota lags behind most other states in the formation of new

TABLE 9.1 Rankings of the Twin Cities Within the 25 Largest Metropolitan Areas

Characteristic	Central Cities	MSA
Poverty rate (1989)	17 (18.5%)	24 (8%)
Population	14 (640,000)	16 (2,464,000)
Population growth 1980-1990	12 (-.2%)	12 (15%)
Central city share of the MSA population	12	
Percentage of population racial minority	24 (20%)	25 (8%)
Percentage of high school dropouts among adults	21 (11%)	24 (7%)
Percentage of adults with 4 years of college	7 (29%)	6 (27%)
Median family income	9 ($33,364)	5 ($43,252)
Violent crime rate	18 (1344)	23 (478)
Married couples as a percentage of all households	17 (36%)	3 (55%)

SOURCE: "The Twin Cities Compared With Other Metro Areas," Appendix 1 in *Keeping the Twin Cities Vital*, Twin Cities Metropolitan Council (1994e), pp. 17-20.

corporations. Most of the state's large manufacturers have programs of plant expansion in other states, while many of their in-state facilities are aging and need replacement (Citizens League, 1994). Business leaders complain that they are discouraged from expanding within the state because of Minnesota's high-tax, high-spend environment and its rigorous regulatory atmosphere. It is difficult to weigh the merits of these complaints, but "business climate" has been an issue for several years.

Although the region is a leader in medical technology companies, it has lost ground in other high-technology sectors, especially computer hardware. Two regional computer firms (Control Data and Cray Research) were decimated by the shift away from mainframes and supercomputers.

■ Growing Central City/Suburban Disparities

Some urbanists argue that the overall economic health of a metropolitan area is directly tied to the health of its central city. Dynamic central cities produce dynamic metropolises; decaying central cities produce decaying metropolises (Rusk, 1993; Ledebur & Barnes, 1992; Savitch, Collins, Sanders, & Markham, 1993). As Table 9.1 suggests, Minneapolis and St. Paul have not decayed nearly as much as have many other central cities, but there is nevertheless reason for concern. The trend for the past 20 years has been toward growing disparities between the two central cities and the surrounding suburbs. It is not as though the region did not try to revitalize the central cities. The two central business districts have been rebuilt over the past two decades, and enormous amounts of public funds have been poured into festival market sites, gentrified historic neighborhoods, downtown redevelopments, and industrial parks in the two cities. Despite these efforts, disparities have grown, especially along three critical measures.

A Sharply Increasing Minority Population

Compared to metropolitan areas of similar or larger size, the Twin Cities do not have a large minority population. The minority population has increased sharply over the past 20 years to nearly a quarter of a million people. The minorities come from four population categories, and no one category dominates. This gives the Twin Cities a much more diverse minority population than is the case in most metropolitan areas. The most profound change came in the Asian population, which by 1990 had become the second largest minority group. This growth largely was the result of an influx of thousands of Hmong refugees from Laos and immigrants from Korea, China, and Vietnam.

As Table 9.2 shows, the suburbs are picking up a larger share of the minority population as time goes on. The concentration of African Americans and Hispanics decreased in the 1980s. In 1990, 15.5% of African Americans lived in census tracts that had black majority populations, down from 25.7% in 1980 (*Minneapolis Star Tribune,* October 4, 1991, p. 3Be). Nevertheless, most minorities still live in the central cities. During these same years, the number of whites in the central cities declined by almost 200,000. Minorities now compose about 20% percent of the two central cities, and they compose a majority of students in the public school systems.

Spreading Poverty

Poverty has gotten worse in the central cities. In 1970, the two central cities had only 19 census tracts with more than 25% of persons living below the poverty line. This expanded to 34 census tracts in 1980 and 51 in 1990. The only census tracts to drop out of this grouping (3 in St. Paul and 2 in Minneapolis) were those in the gentrifying areas in or near the central business districts (Twin Cities Metropolitan Council, 1994c, p. 10-15).

TABLE 9.2 Population Growth by Race

	1970	1980	1990	Change 1970-1990	Change 1980-1990
Central cities					
White	703,000	566,900	512,800	−190,200	−54,100
Minority	41,380	74,286	127,818	86,438	53,532
Suburbs					
White	1,091,612	1,314,317	1,573,859	482,247	259,542
Minority	19,896	42,973	83,965	64,069	40,992
Region					
White	1,874,612	1,881,225	2,086,659	212,047	205,434
Minority	61,276	117,259	211,783	150,507	94,524
Black	32,140	49,270	89,459	57,319	40,189
Asian	4,953	29,970	64,583	59,630	34,613
Hispanic	11,700	21,866	36,716	25,016	14,850
Indian	9,958	15,666	23,340	13,382	7,674
Percentage of minorities in the central cities	67.5	63.4	60.4		

SOURCE: "Trouble at the Core: The Twin Cities Under Stress" (Metropolitan Council Staff report), Twin Cities Metropolitan Council (1992), pp. 51-52; "Population Growth and Residential Development," Appendix 6 in *Keeping the Twin Cities Vital*, Twin Cities Metropolitan Council (1994b), pp. 6-8.

Until 1970, poverty in the Twin Cities was concentrated in four or five main pockets. Today, census tracts with 25% poverty rates extend in an unbroken belt from the East Side of St. Paul roughly parallel to Interstate 94 through the center of Minneapolis to the western boundary of that city. As poverty spreads geographically, it puts pressure on nearby working-class neighborhoods and even some of the inner ring suburbs (*Minneapolis Star Tribune,* October 13, 1992, p. 5). The high concentration of poverty in the central cities is accompanied by a higher concentration of social problems than in the surrounding suburbs. The central cities contain more than 80% of the region's homeless population (Twin Cities Metropolitan Council, 1994d, p. 5.4). They have the highest rates of births to single women, the highest percentage of low-weight births, the highest percentage of mothers with no prenatal care, and the highest rates of crime and violent crime (Twin Cities Metropolitan Council, 1994d, pp. 9, 12, 11, 31-36). Furthermore, on most of these statistics, the current rates are higher than those of 10 or 20 years ago.

TABLE 9.3 Twin Cities Area Job Growth: 1980-1990

	Number of Jobs		
	1980	*1990*	*Change*
Central cities	445,371	450,818	5,447
Inner suburbs	326,760	379,693	52,933
Fully developed area total	772,131	830,511	58,380
Developing area	214,237	374,273	160,036
Rural area	53,643	88,337	34,694
Regional total	1,040,011	1,293,121	253,110

SOURCE: "Job Location," Appendix 9 in *Keeping the Twin Cities Vital*, Twin Cities Metropolitan Council (1994a), pp. 9-16.

Negligible Job Growth in the Central Cities

Although the last 20 years have brought a big influx of immigrants and a tripling of the minority population, there has been negligible job growth in the central cities, accompanied by a significant increase in people below the poverty line. This is shown in Table 9.3. Hundreds of millions of dollars were spent on Minneapolis and St. Paul redevelopment projects in the 1980s, but only 5,447 net new jobs were added in the central cities. The inner suburbs added only 52,933. The bulk of job growth took place in the outlying developing area, where there was a 75% increase in jobs.

Topping off the cities' meager job growth has been a loss of manufacturing jobs. From 1988 to 1992, the State Department of Jobs and Training counted a loss of 11,208 manufacturing jobs in the central cities and a loss of another 10,470 manufacturing jobs in the suburban balance of Hennepin (Minneapolis) and Ramsey (St. Paul) counties (*St. Paul Pioneer Press*, February 4, 1994, p. 8D). This is a critical loss because of the key role that manufacturing jobs historically played in making the city a place of upward mobility for millions of European immigrants and American-born lower-class people. The loss of unskilled production jobs is especially hard on those immigrants and minorities with educational and training deficits.

This migration outward of jobs would not be so devastating if the core area residents could commute easily to the developing areas where employment is growing. Many core area residents, however, do not own cars. Surveys of core neighborhoods indicate that 48% of black, 56% of Native American, and 36% of Asian households lack an automobile (Twin Cities Metropolitan Council, 1992, p. 21). Public transit does not make up for the dearth of cars, because bus service is minimal beyond the major shopping malls.

■ The Great Hopes for Metropolitan Governance

What makes the growing regional disparities especially discouraging for concerned Twin Citians is that steps were taken a quarter of a century ago to forestall urban decline. In 1967, the state legislature created a governing structure for the seven central counties and gave it responsibility for grappling with land use, housing, transit, sewage, and other metropolitan issues. At the top of the regional governing structure was the Metropolitan Council, which, for the next decade, did an outstanding job of tackling those issues. The question today is whether this metropolitan governing structure can provide direction on the threefold issues of regional disparities, suburban sprawl, and positioning the region for global competition.

An Innovative Governing Model

The governing model in the Twin Cities had two fundamental principles. First, it was a bifurcated model that separated metropolitan policy making from program implementation. The centerpiece of the model is the Twin Cities Metropolitan Council, which was responsible for setting policies on transit, water quality, housing, land use, and other issues of metropolitan significance. Those policies were implemented, however, not by the Metropolitan Council but by other governing bodies. This bifurcation of policy making from implementation was established with the Council's first major success, its solution to a metropolitan sewerage crisis. Many growing suburbs had neither a central sewer system nor a central water supply. The State Department of Health reported that nearly half the homes it tested in these suburbs were using drinking water polluted by backyard septic tanks (Harrigan & Johnson, 1978, p. 27). The Federal Housing Administration threatened to stop insuring home mortgages if the homes were not tied into a sewage treatment system. When the state legislature created the Council in 1967, it specifically instructed the Council to find a solution to this problem. The Council recommended the creation of the Metropolitan Waste Control Commission to run a metropolitanwide system of treatment plants and trunk sewer lines. Policy control over where these sewer lines would be built was vested in the Metropolitan Council, giving Council leaders optimism that they could shape future growth of the metropolitan area. They would simply restrict sewer capacity in places where they wanted to inhibit growth.

The second significant feature was a legislature-guided process of incremental fine tuning. Each legislative session saw the Metropolitan Council ask the legislature for authority and funding to deal with a major issue. In 1969, the legislature dealt with the transit issue by putting the Metropolitan Transit Commission under the policy guidance of the Metropolitan Council. Further increments took place in the 1973-1974 session, when the regional governing model was restructured. The Metropolitan Council was given greater control over the Waste Control Commission, the Transit Commission, the Airports Commission, and other regional commissions. A Housing and Redevelopment Authority was created, and this gave the Metropolitan Council the ability to build moderate-income housing in the smaller suburbs that lacked the resources to do so on their own.

Thus, within a decade of its creation, the Metropolitan Council sat atop a complicated system of metropolitan governance. It appointed the members of most metropolitan commissions, had approval power over their capital budgets, and through the Metropolitan Development Guide laid out the policies that these commissions were supposed to implement. Because of its legislatively guided incremental innovation, this governing model avoided the paralyzing conflicts over metropolitan consolidation referenda that failed three-fourths of the time that they were attempted (Marando, 1979). It also avoided the paralyzing weaknesses of the Council of Government (COG) model that was proliferating across the country (Mogulof, 1971). Unlike a COG, the Metropolitan Council had financial independence through its own tax base and its ability to win federal grants. Also unlike a COG, the local governments were not represented in the Metropolitan Council. The 16 Metropolitan Council districts overlapped municipal boundaries and were roughly coterminous with two State Senate districts.

In sum, by 1977 the Twin Cities had created a bifurcated metropolitan governing apparatus characterized by separation of policy making (by the Metropolitan Council) from program administration (the other governments) and a legislatively guided process of incremental innovation. This bifurcated system was very complicated in that different service areas had different implementation structures. Sewerage services were most susceptible to policy direction by the Metropolitan Council because of its control over the Waste Control Commission. Airport services, on the other hand, were least susceptible to policy direction by the Metropolitan Council, because the Metropolitan Airports Commission enjoyed much more autonomy than did the Waste Control Commission.

A Decade of Unprecedented Accomplishments

Not only was this model a structural innovation, it also produced a string of impressive accomplishments in its first decade. It developed a solution for the sewage crisis and a systematic regional approach for solid waste disposal. It forced the affluent suburb of Golden Valley to develop a plan for low- and moderate-income housing. It vetoed plans of the Airports Commission for a second international airport, and it also vetoed plans of the Metropolitan Transit Commission to build a rapid rail transit system. The vetoes over these two huge metropolitan authorities constituted a significant political achievement. No comparable metropolitan agency in the country had ever accomplished such a feat.

Perhaps the Council's most renowned achievement came in 1971. The legislature passed the Council's proposed Fiscal Disparities Act, which provided that the entire metropolitan area, including the central cities, share in the tax base created by new commercial real estate.

Another far-reaching achievement was the creation of a seven-county system of regional parks. The Council achieved bonding authority to purchase land for parks and open spaces. True to its principle of separating policy from administration, the Metropolitan Council issued the bonds but turned the money over to the counties, which actually acquired the land and built the parks. The result today is an extensive system of regional parks in places that most likely would have become shopping centers or residential neighborhoods.

Even more far-reaching was a systematic approach to land planning set up in 1975-1976. In 1976, the Council passed the Development Framework Plan, which divided the region into five development areas—the Central Cities, the Fully Developed Area, the Developing Area, the Rural Area, and the Free Standing Growth Centers. Around the developing area was drawn a Metropolitan Urban Services Area (MUSA) line that roughly paralleled the interstate highways that encircle the Twin Cities. By channeling future growth to the developing area and the freestanding growth centers, the Council hoped to contain suburban sprawl and save local

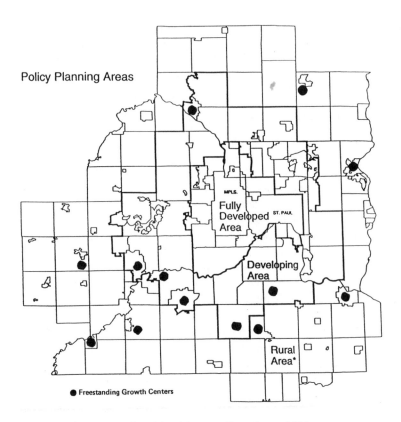

Figure 9.1.   The Twin Cities Metropolitan Area, 1992
SOURCE: Twin Cities Metropolitan Council.

taxpayers from spending $2 billion for public infrastructure over the next two decades (Figure 9.1).

In 1976, the legislature put teeth in this plan by passing the Metropolitan Land Planning Act. This act required all the region's municipalities and townships to submit comprehensive plans to the Council for approval. If a local plan was inconsistent with the growth projected for its municipality by the Development Framework, the Council had authority to hold up the plan's approval. This gave the Council considerable power to pressure local governments into bringing their plans into conformity with the Development Framework. The net result was an elaborate, multiyear process of negotiating back and forth between the Council and local municipalities as they went about the task of implementing the Development Framework. Finally, adding even further strength to the Council's control over land use, the Council in 1976 was given power to review proposed projects of "metropolitan significance" to ensure that those projects were consistent with the Development Framework and other policies of the Metropolitan Development Guide.

Again, consistent with the role bifurcation model, the Council did not implement any of the local development plans. It established the overall metropolitan land planning policy and put limits on the growth that would be permitted in each municipality. Just how and where that growth would occur was left to each municipality to decide for itself using zoning codes, building permits, variances, subdivision requirements, and all the other tools of land management. The municipalities had to use these tools, however, within the confines of overall metropolitan land planning policy. With a few exceptions such as Hawaii, Vermont (Haskell & Price, 1973), and the Coastal Zone Commissions in California (Mogulof, 1975), this was the most systematic attempt to control metropolitan sprawl that the nation had yet seen.

In sum, by the tenth anniversary of the Metropolitan Council in 1977, the bifurcated governance model had gained considerable legitimacy, had evolved an innovative governing structure that fit the desires of regional elites, and had scored an impressive array of accomplishments in carrying out its mission. There were great hopes that this system of metropolitan governance would lead the Twin Cities into a bright future and serve as a model that would be copied by other regions around the country.

■ The Great Hopes Gone Sour

By 1990, those great hopes had gone sour. The Metropolitan Council faced persistent criticism throughout the 1980s from the legislature (Legislative Commission on Metropolitan Governance, 1983), the Legislative Auditor (Legislative Auditor of Minnesota, 1985), the Citizens League (1984), and an editorial writer for the *Minneapolis Star Tribune* with a long history of supporting the Council (Whiting, 1984). Despite its vaunted powers, the Council has played only a bit role in some of the most significant land use decisions over the past 20 years, ranging from the building of a domed stadium (Klobuchar, 1982) that turned out to be a success to a racetrack that ended up in bankruptcy. In between were a World Trade Center for St. Paul, a velodrome for the northern suburbs, the nation's largest shopping mall for the southern suburbs, a state-of-the-art basketball arena for downtown Minneapolis, numerous industrial parks, and numerous festival markets, some of which are in severe financial trouble.

There is no way to know how a strong, resourceful Metropolitan Council would have ruled on any of these developments, because the Council did not try to impose its will on any of them. The more important the project was to big development interests, the less the Council seemed to affect the decisions. The result is that the siting of most of these projects resulted from the traditional politics of land use rather than from a guided land use policy directed by the Development Framework. The racetrack is a good example. In the 1980s, Minnesota voters approved a Constitutional amendment removing any constitutional impediments to pari-mutuel betting. From that point on, it was evident that a racetrack would be built someplace in the metropolitan region. The key contestants for the site were Blaine in the northern suburbs, Woodbury in the eastern suburbs, and Shakopee in the southwestern suburbs. These municipalities competed with proposals for tax increment financing, industrial development revenue bonds, and other enticements. To critics, it was exasperating to see this competition of tax dollars to attract a development that was going to be built someplace in the area even if not a penny in public dollars was spent. There is no evidence that the Metropolitan Council used its powers of metropolitan significance to shape the siting decision in accord with its own metropolitan land use plans. Shakopee finally won the prize, which turned out to be a bit tarnished when the novelty of horse racing eventually wore off and attendance declined. The racetrack went bankrupt in 1993. Horse racing may return to the track, but voters rejected an off track betting amendment in 1994, limiting future prospects for the industry.

Perhaps the single largest land use decision, with the most far-reaching implications on the metropolitan area, was the decision to build the Mall of America on the site of the former baseball and football stadium in the southern suburb of Bloomington. This site had been vacant since the domed stadium opened in downtown Minneapolis. The site was located along the largest strip of commercial real estate in the Twin Cities, and Bloomington was eager to develop the empty space. When constructed, the Mall of America was the largest shopping mall in the nation.

Although the Metropolitan Council appears to have had no influence on the decision to build the Mall of America, it did influence the mall in ways that reduced congestion, increased the amount of retail space, and prevented the construction of a proposed convention center next to the site (personal communication, Roger Israel, August, 1994; personal communication, Kenneth Reddick, August, 1994). The jury is still out on this development's ultimate impact on the Twin Cities, but to date the Mall of America appears to be financially sound. It has attracted tourists to the Twin Cities, and (other than downtown St. Paul) it does not appear to have harmed the previously existing retail areas. It is conceivable that in the long run the region is better off that the Council was bypassed in the Mall of America decision.

In addition to being bypassed on significant land use decisions, there were other signs that the Metropolitan Council was losing its dynamism by the late 1980s. Despite their important decision-making potential, Council members are almost invisible to the general public. The Council found itself in a decade-long irresolvable controversy with the Transit Commission over the management of the bus system, planning for transit policy, and light rail transit. It also came under criticism for permitting expansions in the MUSA line (Alnes, 1992). Although the Metropolitan Urban Services Area had expanded by only 2% since 1986 (personal communication, Kenneth Reddick, August, 1994), the *Minneapolis Star Tribune* accused the Council of becoming an "easy mark for leapfrog developers and tax-base hungry fringe suburbs" (*Minneapolis Star Tribune,* March 5, 1994, p. 14A).

A new governor in 1991 threatened to eliminate the Council if it failed to bring about service efficiencies through the consolidation of local government services. Budget problems forced a 20% reduction in planning staff in 1993. It was clear that another major crisis was brewing for the Metropolitan Council.

■ Regime Change and the Metropolitan Crisis

No government exists in a vacuum. Urban scholars often use the term "regime" (Stone, 1987) for the vague system of leadership ties between public- and private-sector elites. In the late 1960s, when the metropolitan governing model was created, a dynamic metropolitan regime existed that provided powerful support for the Metropolitan Council's approach to metropolitan issues. One of the most serious metropolitan problems today is that regime changes over the past quarter of a century have dissipated important sources of support for the Metropolitan Council. Much of the explanation for the collapse of the bifurcated governing model in the 1990s can be traced to these regime changes. The most important of these involve the legislature, the business community, the political parties, and the failure to create a position of metropolitan leadership.

The Legislature

The legislature historically played a key role by enacting Metropolitan Council initiatives into law. In the early days, when the Council marched from one stunning success to another, Council successes reflected positively on legislators who had in turn supported Council initiatives. It has been a long time, however, since the Council has scored a stunning, highly visible success. The legislature has very little institutional memory of the Council as a dynamic positive force in the Twin Cities, and few legislators are left who can build political support for themselves by pointing to their work on Metropolitan Council initiatives as positive achievements. Of the 201 representatives and senators in the 1993-1994 session, only 2 held office at the creation of the Council, and only 24 held office during the Council's first decade of success (Secretary of State of Minnesota, 1993). Only 12 of these old-timers are from the seven-county region where the Council operates, and of that dozen, none is a prominent advocate of the Metropolitan Council or the regional governing structure.

Contrasting the Metropolitan Council's current standing in the legislature with its standing 20 years earlier, one is reminded of President Kennedy's quip as he absorbed blame for the Bay of Pigs disaster at the beginning of his presidency. "Victory has a hundred fathers," Kennedy supposedly quipped; "defeat is an orphan." The Council sorely needs some victories to end its orphan status and to get some legislators to share in the credit for those victories.

The Business Community

Even more ominous than the changes brought by legislative turnover are the changes in leadership of the Twin Cities business community (Israel, 1993, p. 10). Some of the most important forces for metropolitan reform in the 1960s came from prominent business figures—members of the Pillsbury family and the Dayton family, for example. This generation of business leaders has passed, and the most prominent of the old families no longer control the corporations that bear their names. The change is most symbolic in the case of Dayton-Hudson. As a locally based retailer in the 1960s, Dayton-Hudson had a vested financial interest in the well-being of the Twin Cities, and Donald Dayton was an original appointee to the Council. This financial vested interest in the area is much less the case for Dayton-Hudson now that it has become a national retailing chain. Dayton-Hudson is still a generous contributor to local causes (Oslund, 1994), but rising stars in the corporate hierarchy are no longer tied to this particular metropolitan region. It is no accident that the top officials of major corporations take less interest in metropolitan issues and organizations such as the Citizens League (Byrum, 1994). The new leaders of these corporations have less need to show a local commitment to the Twin Cities than did their predecessors a generation earlier.

The Political Parties

Compounding these leadership problems has been a decline in the ability of the political parties to play a stabilizing role in Twin Cities politics. Increasingly in the late 1980s and the 1990s, the party activists who control party endorsements and party platforms have come from the ideological extremes. This is especially pronounced in the Republican Party, in which pro-life forces dominated the state conventions in 1990 and 1994 and endorsed gubernatorial candidates who were extremely divisive. It is hard to reconcile today's limited-issue Republican leadership with the broad-based Republican leaders who created the Metropolitan Council in 1967. To a lesser degree, the mirror image of the Republican activist appears in the Democratic Farmer Labor Party, in which the most dominant activists are pro-choice, gay rights, and feminist advocates, and people who actively seek to shut down the state's nuclear power plants. (The Democratic Party in Minnesota is known officially as the Democratic Farmer Labor Party, or, more commonly, the DFL.)

It is possible that the ideologues who dominate the two parties may well have the answers to today's global problems, but they exhibit little interest in the metropolitan governing model. For this reason, the parties are in a poorer position today to provide the stabilizing political role that they performed in the 1960s, when the Metropolitan Council was created.

No Position of Areawide Leadership

Finally, the metropolitan governing model deteriorated because there is no highly visible political figure whose own political base depends on coping with the metropolitan issues of central city/suburban disparities, suburban sprawl, and positioning the region to compete in the global economy. In many instances, political leaders would be harmed by trying to develop such a political base. This is certainly true of the governor. As the political parties increasingly marginalize themselves by falling under the sway of ideologically extreme factions, the governor increasingly gets elected through a personalized campaign aiming more at constituencies that cross over party labels than at party loyalties. In this type of electoral campaign, metropolitan issues are either nonexistent or emerge in codewords dealing with crime, jobs, and poor educational achievement. Although reducing metropolitan disparities and positioning the metropolis for the global economy come pretty close to what Paul Peterson called "unitary interests" that would benefit everybody in the region (Peterson, 1981), it is hard to imagine anybody capable of winning the governorship by campaigning on these issues.

The most visible metropolitan positions are the mayors of Minneapolis and St. Paul. For the past 20 years, those offices have been held most of the time by liberal individuals who want to address the issues of urban decline. Nevertheless, their cities compete bitterly for the same limited development dollars, and their political representation in the legislature declined as the Twin Cities fell behind in population growth. For these and other reasons, neither mayor has become a major force on the urban issues addressed here.

Finally, the Metropolitan Council has failed to acquire a position of metropolitan leadership. The Metropolitan Council staff has done a first-rate job of producing planning documents and analyses of the region's demographic and economic trends, but once the Council's first decade of achievement was over, the Metropolitan Council chairperson and members failed to use their positions to dominate metropolitan issues. The problem is that Council members have no political base of their own. They are alternately ignored by the governor and dominated by the governor (Israel, 1993). With the legislative turnover described earlier, they have lost their base of support in the State Legislature. There is a serious fear that the Council has become little more than another state agency competing for the favor of the governor (personal communication, Roger Israel, August, 1994).

■ Tinkering With the System in the 1990s

To recapitulate, the dawn of the 1990s saw several adverse developments that hindered the ability of the metropolitan governing system to cope with the issues of the day. The nature of the most pressing issues had changed from physical development in the 1960s to the much more intractable issues of social deterioration in the 1990s. Economic and demographic change worsened the central city/suburban disparities. Changes in the metropolitan leadership regime diluted the base of support for the Metropolitan Council and its priorities. The Council's authority to cope with urban issues was not expanded. It suffered a decline in federal grants-in-aid monies during the Carter and Reagan Administrations, and the State Legislature did not make up those shortfalls by increasing its own grants or increasing the Council's taxing powers. Most important, the Council failed to develop an independent political base. As an appointive rather than an elective body, it lacks the political clout to confront the governor, key legislators, and sometimes even local officials.

While the Metropolitan Council was losing political power, the counties were gaining influence. In Minnesota, it is the counties that have primary responsibility for social service and public health programs. As their budgets grew and as the county governments modernized themselves, they became formidable forces for defending their turf against metropolitan-level policy making. In 1993, the counties proposed legislation that would have reduced the Council's influence even further (*Minneapolis Star Tribune,* January 6, 1994, p. 18A).

In the 1993-1994 sessions of the legislature, two developments brought these issues to a head. The first was transportation, and the second was a package of bills authored by a persistent Minneapolis legislator who wanted to address the growing central city/suburban disparities.

Transportation and Metropolitan Reform

Transportation had been a sticky issue for many years. Although the Metropolitan Council effectively had vetoed the Transit Commission's (MTC) plans for a rapid rail line in 1974, rail advocates came back in the 1980s with proposals for a light rail transit (LRT) system. With the MTC's planning role reduced and the Metropolitan Council opposed to rail-based transit, LRT advocates complained that overall transportation policy was being neglected. The legislature responded in 1989 by creating a Regional Transit Board (RTB) that was placed under the Metropolitan Council and over the MTC. This structural change did not resolve the transit conflicts. It replaced a two-level system with three levels in which the MTC was primarily an operating agency nominally responsive to both the RTB, which advocated rail transit, and the Metropolitan Council, which historically opposed rail transit and had at best become a skeptical and cautious supporter of it. Muddying the issue further, Hennepin County and Ramsey County approved plans for an LRT system and asked the legislature in the 1993-1994 session to fund a plan that essentially would follow the same rapid rail paths that the Metropolitan Council had vetoed in 1974. The counties funded a "grassroots" group called Minnesotans for Light Rail Transit to build public support for LRT and lobby the legislature (Dornfield, 1994).

While these transit conflicts were going on in the background, a highly visible transit issue undermined both the RTB and the MTC. The MTC and later the RTB had operated Metro Mobility, which provided on-demand transportation service for disabled people. In 1993, the RTB awarded the Metro Mobility contract to a new service provider, which promptly ran into snags with its dispatching system. Service deteriorated and, as complaints mounted, the governor responded by calling out the National Guard to drive the Metro Mobility vans until a solution could be found. Metro Mobility riders sued over their poor service and eventually received a $1.3 million settlement from the RTB (*St. Paul Pioneer Press,* March 8, 1994, p. 1A). As if this were not enough to undermine the transit system, the MTC board fired the MTC's chief administrator, whom the governor viewed as having done an excellent job of rebuilding bus ridership. The governor responded by trying to force the MTC to hire back the administrator it had just fired.

These events turned out to be the coup de grace for the bifurcated model of metropolitan transit. On paper, the MTC board ultimately was accountable to the governor, because the MTC board was appointed by the RTB, which was in turn appointed by the Metropolitan Council, which was in turn appointed by the governor. From one perspective, the governor's attempt to reinstate the fired transit chief was an unwarranted intrusion into the operational affairs of a metropolitan agency. From the opposite perspective, the MTC was an agency out of control, impervious to democratic oversight. There was no way to look at the transit crisis of 1993-1994 without concluding that the traditional bifurcated model of governance was in a shambles.

Thus, the Metro Mobility crisis and conflicts over LRT set the stage for metropolitan reform in 1993-1994. Issues were raised that were not resolvable within the bifurcated model.

A Metropolitan Legislative Package

The second thing provoking a metropolitan restructuring in 1993-1994 was a package of legislative bills named after its author, Minneapolis State Representative Myron Orfield. Orfield addressed the growing central city/ suburban disparities with a package of bills that would provide for

- an elected Metropolitan Council,
- mandating the Council to prescribe low- and moderate-income housing goals for each suburb,
- giving the Council authority to enforce its low- and moderate-income housing goals by denying sewers and highway extensions to suburbs out of compliance, and
- a housing reinvestment fund providing that for each new home valued above $150,000, a percentage of the excess tax base valuation would go into a housing reinvestment fund that would be used to increase the supply of low- and moderate-income housing.

Orfield calculated that three-fourths of the suburbs would benefit from the reinvestment fund and only one-fourth of the suburbs would pay more into the fund than they got out of it. That one-fourth included the most affluent suburbs in the region, and they bitterly opposed the proposals. Orfield spent the better part of 2 years testifying at hearings and traveling throughout the region to address any group that was willing discuss his proposals and to look at his many charts on growing central city/suburban disparities.

Incremental Tinkering in 1994

All these proposals were put before the 1993-1994 sessions of the legislature. Orfield's housing bills passed the Democrat-controlled legislature in 1993 but were vetoed by the Republican governor. The legislature put off a decision on funding for the LRT proposals. In 1994, the bill for an elected Metropolitan Council failed by a single vote in the House and only five votes in the Senate.

The most significant action that did take place was a major change in the bifurcation governing model, which had been in place since 1967. The legislature abolished both the RTB and the MTC and turned all transit operations and planning over to the Metropolitan Council. Likewise, the legislature abolished the Metropolitan Waste Control Commission and gave its operations to the Council as well. Because the Council would now be a major operating agency, the legislature created the position of Regional Administrator. The person in that position was expected to serve as a city manager for Metropolitan Governance. Finally, to tighten accountability over metropolitan governance, the governor was given authority to dismiss as well as appoint members of the Metropolitan Council.

■ Conclusion

The 1994 changes in the metropolitan governing model were significant. They discarded the bifurcation model for transit and waste control operations by, for the first time, making the Metropolitan Council responsible for both policy planning and operations. It is not likely that the other metropolitan commissions (Sports Facilities and Airports) will also lose their operating responsibilities, unless a crisis emerges similar to the one that brought the governor directly into the dispute over transit.

The restructuring of 1994 also suggests that a new metropolitan leadership regime might be emerging. Both the Orfield package and strengthening the Metropolitan Council received solid support from the daily newspapers and the Association of Metropolitan Municipalities. Even suburban Republican legislators admitted the need to deal with the regional disparities and proposed a milder version of Orfield's package that was dubbed "Orfield Light" by the press. Only the counties and a small number of affluent suburbs stayed consistently in opposition. If this new metropolitan leadership regime stays in place as the metropolitan issues return in future years, it will greatly enhance the prospects for action.

The issues that provoked so much activity in 1993-1994 have not been resolved. Despite the fact that transit operations have now been centralized in an agency that historically has been cool to rail forms of transit, the proposal to build an LRT system still has many vocal and influential advocates, and the guerrilla warfare over LRT will probably continue.

The Orfield package to alleviate housing disparities between the central cities and the suburbs also will remain on the agenda. Many elites in the region seem to believe not only that the disparities are growing but also that they pose a threat to the overall quality of life and the region's economic stature.

The legislature failed to take the single biggest step it could have taken to resolve these issues—providing for an elected Metropolitan Council in order to create a position of metropolitan leadership with its own political base. Instead, the 1994 legislature increased the dependence of Council members on the governor. This means that the Council's ability to address the growing central city/suburban disparities will depend primarily on the goodwill of the governor. If the governor favors taking the political risks to address those issues and is willing to back up the Council's efforts along those lines, then it is conceivable that considerable progress could be made. In the long run, however, tying the Council so closely to the governor makes action on metropolitan issues dependent on the whim of the governor, who has little to gain politically from involving himself or herself in the divisive issues of regional disparities and growth control. There is a great danger that the Council will become simply another state agency. The pressure on the Metropolitan Council members, like leaders of state agencies, will be to avoid doing things that might cause problems for the governor, but action on regional disparities and suburban sprawl inherently causes problems for the governor.

In the last analysis, the key issue is still political. Alleviating regional disparities, containing sprawl, and positioning the region for global competition cannot be achieved by a bureaucratic agency. It can be done only through leadership that has its own political base of support. Rather than creating such a base of support by providing for an elected Metropolitan Council, Minnesota in 1994 decided instead to make the Metropolitan Council's political strength dependent on the governor. This might work, but it probably will not.

REFERENCES

Alnes, S. (1992, December 15). Met Council faces push for new roles amid calls for more power in old ones. *Minnesota Journal, 9*(12), pp. 1, 4.

Byrum, O. (1994, April). Seminar on metropolitan organization, Hamline University, St. Paul, MN.

Citizens League. (1984). *The Metro Council: Narrowing the agenda and raising the stakes.* Minneapolis, MN: Author.

Citizens League. (1994). *Call for membership applications to Global Positioning Task Force.* Minneapolis, MN: Author.

Dornfield, S. (1994, July 18). A lot of public bucks go into backing "grass-roots" citizens group for LRT. *St. Paul Pioneer Press,* p. 6A.

Harrigan, J. J., & Johnson, W. C. (1978). *Governing the Twin Cities region: The Metropolitan Council in comparative perspective.* Minneapolis: The University of Minnesota Press.

Haskell, E., & Price, V. S. (1973). *State environmental management.* New York: Praeger.

Israel, R. (1993). *Twin Cities intergovernmental relations: The changing landscape.* Unpublished manuscript.

Klobuchar, A. (1982). *Uncovering the dome.* Prospect Heights, IL: Waveland.

Ledebur, L. C., & Barnes, W. R. (1992). *Metropolitan disparities and economic growth* (Research report). Washington, DC: National League of Cities.

Legislative Auditor of Minnesota. (1985). *Metropolitan Council.* St. Paul, MN: Author.

Legislative Commission on Metropolitan Governance. (1983). *Report of the Legislative Commission on Metropolitan Governance.* St. Paul: Minnesota Legislature.

Marando, V. L. (1979). City-county consolidation: Reform, regionalism, referenda, and requiem. *Western Political Science Quarterly, 32*(4), 409-422.

Mogulof, M. B. (1971). *Governing metropolitan areas.* Washington, DC: Urban Institute.

Mogulof, M. B. (1975). *Saving the coast: California's experiment in inter-government land use regulation.* Lexington, MA: Lexington Books.

Mollenkopf, J. H. (1983). *The contested city.* Princeton, NJ: Princeton University Press.

Oslund, J. J. (1994, April 4). Fewer firms set fixed goals for corporate giving. *Minneapolis Star Tribune,* p. 1D.

Peterson, P. (1981). *City limits.* Chicago: University of Chicago Press.

Rusk, D. (1993). *Cities without suburbs.* Washington, DC: The Woodrow Wilson Center Press.

Savitch, H. V., Collins, D., Sanders, D., & Markham, J. P. (1993). Ties that bind: Central cities, suburbs, and the new metropolitan region. *Economic Development Quarterly, 7*(4), 341-357.

Secretary of State of Minnesota. (1993). *Legislative Manual of Minnesota: 1993-94.* St. Paul, MN: Author.

Sternberg, B. (1993, April 25). Region's residents feel tug of Twin Cities. *Minneapolis Star Tribune,* p. 12A.

Stone, C. N. (1987). Summing up: Urban regimes, development policy, and political arrangements. In C. N. Stone & H. Sanders (Eds.), *The politics of urban development* (pp. 269-290). Lawrence: University of Kansas Press.

Twin Cities Metropolitan Council. (1992). *Trouble at the core: The Twin Cities under stress* (Staff report). St. Paul: Author.

Twin Cities Metropolitan Council. (1994a). Job location. Appendix 9 in *Keeping the Twin Cities vital.* St. Paul: Author.

Twin Cities Metropolitan Council. (1994b). Population growth and residential development. Appendix 6 in *Keeping the Twin Cities vital.* St. Paul: Author.

Twin Cities Metropolitan Council. (1994c). Poverty data. Appendix 10 in *Keeping the Twin Cities vital.* St. Paul: Author.

Twin Cities Metropolitan Council. (1994d). Social indicators. Appendix 5 in *Keeping the Twin Cities vital.* St. Paul: Author.

Twin Cities Metropolitan Council. (1994e). The Twin Cities compared with other metro areas. Appendix 1 in *Keeping the Twin Cities vital.* St. Paul: Author.

U.S. Bureau of the Census. (1993). *Statistical Abstract of the United States: 1992.* Washington, DC: Government Printing Office.

Whiting, C. C. (1984). Twin Cities Metro Council: Heading for a fall? *Planning, 50*(3), 4.

10 Jacksonville: Consolidation and Regional Governance

BERT SWANSON

■ Introduction

The modern metropolis is the dynamic center of the nation's economic and social life. Yet a sense of crisis and doom pervades the "eclipse of community," caused by the processes of urbanization, industrialization, and bureaucratization (Stien, 1960). The settlement pattern suggests a "schizoid polity" (Greer, 1963) with extreme social, economic, and racial disparities between lower-income families residing in the large central city and the more affluent who have moved to the suburbs in a process of "white flight." Greer characterized these evolving patterns as a dichotomy between the residents of the central city and suburbanites, a dichotomy that has led to disorder and conflict in the metropolis. He implied that the metropolis had developed a "multiple personality," with inhabitants withdrawing from reality through illogical—even bizarre and sometimes delusional—behavior, creating a fantasy world with intellectual and emotional deterioration, suffering disturbances in mood and behavior.

Although most analysts have treated the community as autonomous and without connection to the conditions, political trends, and socioeconomic decisions made elsewhere, some local leaders were in fact coordinating their actions with external state, regional, and national leaders. The modern metropolis does not respect the formal legal boundaries of the past; informal leaders provide direction of public affairs in the "megalopolis" from Boston to Washington, D.C. (Miller, 1975). Brown (1978, p. 256) suggested that "The main 'engine' of local politics is not the polity within, but the metropolis without." Thus, the urban polity comprises a

229

mosaic of formal and informal, highly differentiated decision-making settings and power arrangements, attempting to resolve distinctive, yet common, if not similar, community problems.

Reformers and their political scientist advisers have for several decades characterized the politics of the metropolis as "balkanized" and "crazy-quilt," as an "organized chaos" of ineffective governments. They formulated guidelines for unifying metropolitan governance by using one government, electing only the most important policy officials, eliminating the tradition of separation of powers, separating administration from politics, and providing a hierarchical command structure (Anderson, 1925). Application of these principles facilitated the elimination of more than half of the nation's 200,000 local governments. Reformers advanced the need for a "gargantuan" single metropolitan government or at least the establishment of a regional superstructure.

Jacksonville is an example of a single-tier or full consolidated metropolitan government that merged the central city of Jacksonville with Duval County. The preconsolidation governments of Jacksonville and Duval County differed in major ways as to structure and function. The city maximized the "checks and balances" of authority. Its legislative body consisted of nine council members residing in districts but elected by the citywide constituency. The executive functions were carried out by a five-member elected commission with administrative responsibility for specific departments. The mayor was one of the commissioners, with ceremonial functions. In addition, voters elected a recorder, municipal judge, treasurer, and tax assessor. The city carried out a full range of municipal functions, including traditional services and meeting such regional needs as an airport, a coliseum and auditorium, a radio station, the Gator Bowl, and baseball parks. The city also owned and operated the electric utility that served most customers throughout the county. The public service needs of the residents of Jacksonville were served primarily by its municipal government, which even operated its own jail, library system, elections process, and tax assessment system.

Duval County, on the other hand, relied primarily upon a five-member elected commission. The county had no chief executive, and most administrative functions were carried out by elected constitutional officers of the state of Florida, including a sheriff, tax assessor, tax collector, and supervisor of election. The county was responsible for the unincorporated area. Four small towns had been established, as well as a separately elected school board and several special districts with appointed boards responsible for port facilities, hospitals, and expressways.

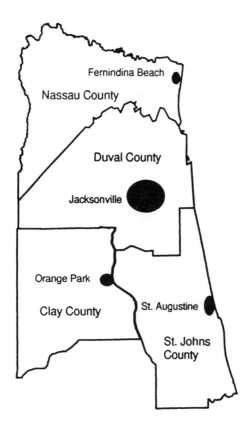

Figure 10.1. The Jacksonville Metropolitan Statistical Area

Jacksonville-Duval has become the key county in the Metropolitan Statistical Area (MSA), which includes the counties of Duval, Clay, Nassau, and St. Johns (see Figure 10.1). Most activities of the consolidated government remain confined within Duval County. The one formal mechanism to facilitate intergovernmental coordination in the metropolis is the Metropolitan Planning Organization (MPO), which addresses the transportation needs of the region (Comprehensive Plan 2010, 1990). Jacksonville dominates the MPO, as all decision makers are from Jacksonville, along with several state agencies' representatives. The state of

Florida has established several regional agencies to regulate the use of water resources (the St. Johns River Management District) and review land use in state-mandated local comprehensive plans to determine their consistency with state and regional plans (the Northeast Florida Regional Planning Council). The Jacksonville municipally owned electric utility (JEA) has evolved into an important regional player, serving customers in the adjoining counties.

The relationships between the four counties in the MSA are mixed. In general, the local governments are cooperative, if not well coordinated; however, there is competition for economic development. Jacksonville lost the prestigious national PGA golf headquarters to St. Johns County by siting it on environmentally sensitive land. Conflicts have arisen over siting LULU (locally unacceptable land uses). Jacksonville was unable to site its new landfill on the boundary with St. Johns County and was forced to relocate and compensate adjoining Baker County.

Jacksonville's domination of the regional economy of northeast Florida has been enhanced by efforts of the Jacksonville Chamber of Commerce to promote its economic-oriented agenda. The state mandated that all localities prepare comprehensive land-use plans that are consistent with the state and regional (NEFRPC) plans. Failure to manage growth and provide public infrastructure to meet population growth could result in a state-ordered "moratorium." Given increasing community diversity, the mayor provided an opportunity for one thousand citizens to engage in a "visions" (City of Jacksonville, 1992) project. By 1994, the consolidated government had created seven district advisory committees to discuss and react to how city projects may affect neighborhoods.

■ Perceptions and Prescriptions of Jacksonville Reformers (The Intervention)

The Jacksonville reformers of the 1960s were "centralists" advocating merger of city and county governments to modernize and professionalize, as well as put whole, the metropolity, which had begun to fragment with the rapid socioeconomic changes associated with burgeoning suburban growth. Local government (city and county) had failed to meet the practical test of responsiveness and adaptability in an orderly and planned way. Reformers proclaimed a high degree of chaos in the public domain, stating that "far too often petty jealousies, selfish interests, delusory 'traditions' and public apathy have been used to calcify a faltering local government system" (Local Government Study Commission of Duval County [LGSCDC], 1966, p. 9). The prevailing sense of the metropolis at the time was that of the Jacksonville SMSA. Jacksonville had long been the dominant center of the regional economy, based on financial, industrial, and military activity.

The preconsolidated governments had responded differently to the potential regional nature of public policy and services. Duval County focused on the three traditional functions of operating the justice system, maintaining farm-to-market roads, and transacting the state of Florida's business, primarily collecting state taxes. The county established a hospital authority, to address the problems of the indigent, and an air improvement authority. The City of Jacksonville provided a variety of urban public services, including a municipally owned electric power system, a regional airport, and sports facilities, all of which were operated as departments within the general city government. Prior to consolidation, the state, through the urging and approval of the Duval legislative delegation, established several Independent authorities. The state established the Jacksonville Expressway Authority to coordinate and build areawide highways. The Port Authority assumed responsibility for the municipal docks and was governed by a seven-member board, five members appointed by the governor and one each by the city and the county. The City of Jacksonville established departments for airports serving regional needs and for housing to publicly support those in need of shelter in municipalities. The city and county had created a joint, but underfunded, areawide Plan Board to take advantage of expanding federal grants.

The advocates of consolidation identified a number of specific problems that they believed required a restructured government: disaccredited high schools; high degrees of water and air pollution; higher crime rates in the city; a high degree of property deterioration (slums); poor land-use planning and zoning; high comparative costs of governmental service; lack of public confidence in local government (grand jury findings); low voter registration; slowing economic growth; traffic congestion; comparatively low wages (coinciding with a large percentage of workers in low-skill jobs); inadequate sewer facilities (areawide), water facilities, fire protection, garbage collection, and libraries (unincorporated area); and racial unrest (LGSCDC, 1966, p. 6). They discerned the significant differentials of available public goods and services for those living in the city versus those living in the unincorporated area.

The reformers accepted and articulated four textbook principles, which guided centralists' proposals of governmental restructuring. These principles were economic efficiency, administrative effectiveness, political accountability, and socioeconomic and political equity (Advisory Commission on Intergovernmental Relations [ACIR], 1974). Many of these ideas are based on presumptions of political behavior that are not well grounded in theory but are promoted to convince practitioners and citizens alike to support the need to improve the governance of the metropolity. The reformers marshaled the prevailing theories and empirical evidence about local political systems to provide an intellectual patina for their largely intuitive desire to reduce fragmentation of local authority by reducing the number of governmental units.

The four principles are potentially contradictory given the diversity of needs and preferences of competing socioeconomic and political interests within the metropolis. For example, achieving accountability, given a low level of knowledge or involvement of electorate in a highly centralized polity, is likely to diminish effectiveness and efficiency. Systematic follow-up research has not revealed the expected economies (Morgan & Pelissero, 1980). Achieving efficiency tends to overwhelm concerns for equity between those in the central city and those in the suburbs. Urban reformers tend to emphasize effectiveness and efficiency, with some reference to accountability (with better representation of minorities on the City Council), but with virtually no reference to the equity and the needs of lower-income families.

In 1967, reformers of Jacksonville successfully persuaded two-thirds of the electorate in the city (40% African Americans) and the suburbanites (virtually all white) to adopt a single-tier consolidation of city and county governments. The Jacksonville metropolis appeared to be splintering in the 1960s, with 60% of the population living in the urban fringe. Several civil rights demonstrations had resulted in riots and threatened the private investments of the financial sector, especially in the downtown area. Confidence in government declined as a dozen public officials were indicted for malfeasance in office. At the same time, reformers blamed the general governments of both the city and the county for the failure to resolve such urban problems as disaccredited high schools (operated by separate school districts), dysfunctional sewerage treatment (installed by private developers in the suburbs), traffic congestion, increasing rates of crime, inadequate storm water drainage, and increasing air and water pollution.

Consolidation was not a form of tax revolt, as the reformers and their supporters believed the new government would provide more public services with essentially the same level of funding. They also expected to participate in the available Great Society program of federal grants. The reformers were willing to increase taxes, if need be; more than 100,000 citizens signed a petition to increase property assessments in order to improve the schools in this rapidly growing metropolis. The voters approved a $35 million bond measure to build additional schools to relieve the pressure from double sessions.

While the reformers compiled a vast array of data and analysis about the increasing costs of the inefficient governments of the city of Jacksonville and Duval County, they relied more heavily on their own experiences and frustrations with the "ineffective" and "unaccountable" performance of local governance. For example, Yates, chair of the merger study commission and former head of Southern Bell, was exasperated by the lack of community planning, which he saw as necessary for his company to respond adequately to residential growth in the suburbs. Other reformers sensed that the "powers that be" were not sufficiently interested in the prospects for growth, preferring instead to control the pace and direction of economic development.

Still other civic leaders became concerned about the turbulent problems associated with an expanding and disadvantaged black population. The community's residential and school segregation and patterns of racism seemed to be on a collision course, beyond the control of public officials. Mayor Burns's response to civil rights demonstrations by young African Americans was harsh and unaccommodating as he prepared for his successful gubernatorial campaign. Businesspeople with large investments in the downtown area, which was surrounded by the black community, identified racial unrest as a problem that was getting worse. Although race was a latent issue, some white leaders understood the need to dilute the influence of black inner-city residents. The struggle for power within the black community was changed in favor of those willing to work with white reformers in the creation of the new government.

The LGSCDC (1966, p. 17) asserted the needs of the total community: "Whether we live in the City or outside the City, Jacksonville is, for all practical purposes, 'our' City. If we are to prosper as an economic area, as a community of the future, as individuals in pursuit of our goals in life, we must ensure that our core city is viable and able to speak to the world as a living testimony of our accomplishments." The reformers believed that Jacksonville was one community requiring one government; that is, the residents in the inner city shared a common interest, or ought to, with those in the suburbs. They firmly believed that a single, big, central government could and would aggregate the common interests of all. Consolidation would not only unite the community but also generate long-term economies of scale, thereby keeping the cost of government low and under control. They emphasized the merits of the handful of mergers that had been adopted, especially in Nashville. They explicitly rejected, however, the two-tier approach of Miami-Dade.

To correct conditions, they preferred to increase the delivery of public goods and services, most especially to the suburbs. They firmly believed that this could best be done by improving the managerial capacity of government, which meant upgrading the professional skills of the bureaucracy. The reformers were highly critical about the lack of land-use planning, as both the city and county had their own planning boards, each with different notions about planning. The city essentially relied on an obsolete set of zoning categories, while the County had "at best a makeshift arrangement with the bulk of the unincorporated area being classified as agricultural with no restrictions" (LGSCDC, 1966, p. 141). Subdivisions were located in the agricultural zone without having been rezoned. Although joint countywide or regional planning was established in 1961, it was underfunded and understaffed.

The reformers believed that the prevailing pattern of governance confused the electorate about who was responsible for the low and declining quality of life. The public had become "apathetic" as it was faced with multiple governments that required many electoral choices. There were 75 elected officials, which reformers proposed reducing to 31; they succeeded in eliminating only 11. The reformers sought to pinpoint responsibility with a shortened ballot and a districted City Council, with approximately 25,000 residents per district. The districts would "ensure" the election of three or four black council members, but suburbanites would gain control of the council by having two-thirds of the districts.

Opponents of the reforms, including public officials and leaders of the Democratic Party, proclaimed consolidation as the specter of big government, one saying it was a "diabolical plot" to "establish a dictatorial power structure by the economic elites" (Martin, 1968, p. 96). The opponents were able to prevail upon the State Legislative Delegation, after an intense struggle between the reformers and the old guard to modify the blueprint design. More specifically, the constitutional officers, such as the sheriff, tax collector, tax assessor, and supervisor of elections, continued to be separately elected. Most department and division head appointments were subject to council confirmation. The elected Civil Service Board was retained, but voters approved an appointed board in 1994. The four small towns, with approximately 20,000 residents, were given the right to opt out of the consolidated government, which they did. By 1993, the small-town voters had strongly approved a straw ballot to create a new county, thereby completely exiting the consolidated government. The number of council districts was reduced from 21 to 14. A set of independent authorities—electric (JEA), transportation (JTA), port (JPA), downtown (JDDA), housing (DCHFA), and sports (JSDA)—remained or were established. They operate beyond the direct, day to day control of the general consolidated government. Nor was the new consolidated government given the degree of "home rule" envisaged by the reformers, as the State Legislative Delegation continued to exercise enormous authority over the city.

Consolidation eliminated the legal distinction between the city and the county, which provided selected public services in five municipalities and minimal services for the unincorporated area outside those municipalities. Consolidation was expected to reduce the distinctiveness of the ever splintering city from the suburban electorate, as voters would elect a common set of public officials with the authority to make coherent policies for the entire community and participate in a political process designed to produce a single, comprehensive polity. The single chief executive, joined by a single legislative body of 19 council members (14 from districts and 5 at large), would speak with one voice for the governed. Table 10.1 summarizes the changes established by the consolidated government.

The blueprint for reform, adapting key elements of "scientific management" of public administration, would centralize governmental authority in one unit, responsible for governing the entire community. It emphasized structural change, assuming that that was sufficient to build a metropolity. Reformers preferred an inarticulate ideology of the pragmatic administrative state rather than an articulate quest for community (Swanson, 1970, p. 134). Nor did they articulate the obvious racial and socioeconomic disparities separating residents of the inner city from those of the suburbs. They assumed that consolidation would modify the existing fiscal disparities of differential tax burdens and benefits. They ignored the facts that citizens in a large metropolis generally prefer control over the level of services and that their interests vary considerably.

TABLE 10.1   Regional Functions: Before and After Consolidation

Before	After
County	*Consolidated government*
General government	Single city/county general government except small towns
Hospital	
Air improvement	Independent authorities:
Expressway	JEA (electric)
	JTA (transit)
City	JPA (air and seaport)
General government	JDDA (downtown development)
Airport	JSA (sports development)
Library	
Electric	

SOURCE: *The Blueprint for Improvement*, Local Government Study Commission of Duval County (1966); annual budget summary (1993).

■ Changing Socioeconomic, Political, and Power Profiles of the Jacksonville Metropolis (Pre- and Post-Intervention Consequences)

The achievements of consolidation will be judged against the expectations of reformers as well as by selective contrasts of socioeconomic and political profiles (Swanson & Swanson, 1979) of some 30 years of reform activity designed to centralize government and unify a very divided community. Not all possible consequences of the restructured government and polity will be discussed here, nor will all recorded changes be attributed to consolidation. This presentation will examine the extent to which the dichotomous nature of this fragmented, if not schizoid, polity has been reduced, has increased, or has remained the same. Did the restructured government reunite the polity that enhances regional policy making?

Metropolitan Social and Economic Disparities

The ACIR (1965) identified socioeconomic disparities that reflect the potential political conflicts between central cities and their surrounding suburbs as they compete for shared tax revenues, financial assistance, welfare programs, and highway funding. They confront one another over double taxation, land-use planning and zoning, and the city's move to annex suburban territory. During the first 25 years of the reformed government, residential settlement patterns continued to show dispersal. Nor has there been much change in socioeconomic status and racial disparities.

At the beginning of the 20th century, Jacksonville was the largest city in the state of Florida. Its population reached a temporary peak at 200,000, and the city did not keep pace with the growth of the state. The city declined while the suburbs grew rapidly in the 1950s. Reformers optimistically projected that the city would again be the largest in the state, with an estimated 885,000 persons by 1980. By 1990, the population of Jacksonville had reached only 672,971 persons. Jacksonville is the center of an MSA of a four-county area of 906,727 persons, with 84% of the labor force working in Duval. More recent estimates are that the city will reach a population of approximately 827,000 by the year 2010. Jacksonville is expected to become a smaller proportion of the MSA, which is projected to have a population of 1,141,000 by the year 2015.

Population growth did not take hold until the 1980s, when business firms discovered a favorable business climate. Early population estimates reflected more the aspirations of economic leaders than reality. Only 20% of residents now live in the area that once was the old city. Of those, two-thirds are African Americans. Nearly two-thirds of the residents of the community are "movers." A quarter of the population has relocated to Jacksonville from outside the metropolitan area. Two-thirds of these have moved to the area from other parts of the South. Virtually all the newcomers are whites.

The population in the former city continued to decline, to approximately 120,000 by 1990, as urban redevelopment programs demolished deteriorated residential buildings to make way for commercial and public development. All the growth has occurred in the former suburbs. More than half (57.1%) of residents of the City of Jacksonville were African American at the beginning of the 20th century. Although their numbers grew steadily, their proportion of the total population declined until 1950. By 1960, however, they were well on their way to becoming the majority in the former central city (National Advisory Commission on Civil Disorders, 1968, p. 391). The merger of city and county reduced their proportion to one quarter of the total, where it remained. By 1990, nearly 90% of African Americans lived in the inner city, but only 40% lived in the old city: Many moved into the former suburbs, tending to settle in what became predominantly black neighborhoods. African Americans composed well over half (57.3%) of the population of the former city by 1990.

Jacksonville is one of the most segregated metropolitan areas in the country. Karl and Alma Taeuber (1965) reported the segregation index as 96.9 by 1960 (i.e., 97 out of every 100 black households would have to be moved to achieve an "integrated" community). Consolidation should have little or no effect on this measure of segregation. All public facilities were segregated until the 1960s. A series of federal court orders desegregated schools and recreation centers, including public swimming pools, golf courses, and the like. The school district was among the half dozen most segregated in the country, with a segregation index of 92 for students and 96 for staff (Farley & Taeuber, 1974).

There were significant differences in socioeconomic status between those who lived in the inner city and those in the suburbs. Nearly half the residents of the city over 25 years of age had no more than a grade school education, whereas half of the suburbanites were high school graduates or had received some college education. Nearly 60% of the city residents in the workforce were blue collar, compared to 40% in the suburbs. The preponderance of white-collar workers lived in the suburbs. As one might expect, given the significant differences in educational attainment and occupational status, family income was strikingly different. The pattern in the inner city was one of a pyramid, with a large base of low-income residents, whereas family income in the suburbs could be represented by a diamond shape, with the largest proportion in the middle. The proportion of city families in the lowest quintile of income was three times that of the suburbs. As one might expect, the proportion of families in poverty was higher in the city, twice as high as in the suburbs. Nearly a third of the families in the city earned less than $3,000, compared to 15% in the suburbs.

Consolidation eliminated the former city of Jacksonville and statistically combined the populations of the inner city and suburbs into one reporting unit. It did little, however, to change socioeconomic status differentials of the residents in the two areas. The people in the "urban core" in 1990 continued to have the highest incidence of poverty and the highest proportion of female-headed households. They also had the lowest level of educational attainment, median family income, employment, and skilled jobs in the community.

There were overwhelming racial disparities between black and white residents prior to consolidation (1960). Only a third of black inner-city residents had attained an education beyond grade school, compared to two-thirds of the suburban whites. Sixty percent of suburban whites were employed in white-collar jobs, compared to less than 13% of inner-city blacks. The impact of these two factors, education and occupation, placed half the black residents of the city in the lowest quintiles of family income, compared to 20% of the whites. Less than 2% of the city's black residents were in the highest quintiles, compared to 13% of the whites. Half the inner-city African Americans were considered to be in poverty, compared to 20% of the whites.

Consolidation structurally merged the two governments (city and county) and combined the socioeconomic characteristics of those living in the inner city with those in the suburbs. Taking into account the entire county, black-white family income disparities were reduced statistically through consolidation, but the reform did little to change the actual racial status disparities. By 1990, African Americans still remained predominantly among the lower-income families, as some 37% were in the lowest quintile, compared to 11% of white families. Less than 3% of black families were in the highest quintile, compared to 11% of white families. The socioeconomic status of whites in the former city declined significantly by 1990, with 70% in the lowest two quintiles, compared to 43% in 1960.

Shifting Public Finance

One might expect a rapidly growing community with a reformed government to experience a significant shift in public finance. The economic base of Jacksonville has diversified during the past 25 years. Public officials dramatically rotated their reliance on various revenue streams away from property taxes and toward user charges. The merger of the two general governments, city and county, reduced double taxation for those residing in the former city. The sense of fiscal disparities continues, however, in the allocation and distribution of public goods and services.

Prior to consolidation, the economy of Jacksonville was concentrated in retail, manufacturing, and service activities. These major components of the labor force were similar for the city and the suburbs. About half the metropolitan labor force resided in the city, but the overwhelming amount of business activity occurred in the city. For example, the dollar amounts of transactions by city businesses were higher than in county businesses, as follows: ten times as much wholesale trade, five times the level of service receipts, three times as much manufacturing, and twice the retail sales. The total net buying income of city residents, however, was considerably below that of suburbanites.

By the 1990s, the economy had diversified considerably. Although service, retailing, and manufacturing continued to be leading business activities, finance, transportation, and wholesale activities had tripled since the 1960s. Government employees—federal, state, and local—numbered about 50,000, 15% percent of the civilian local labor economy. The armed forces accounted for another 25,000.

New downtown skyscrapers—American Heritage (insurance), Barnett (banking), and Southern Bell (telephonic communications)—provided a sense of a boisterous economy. When American Express chose to locate in Jacksonville, the company attributed its decision to the ease of dealing with the consolidated city. The "miracle" of obtaining an NFL franchise bolstered Jacksonville and the northeastern region of Florida. Not all efforts to "grow" the economy, however, have been successful. The Off-Shore Power Systems (OPS) plan to build floating nuclear power plants produced by Westinghouse-Tenneco failed. The major shipyard closed in 1990, and several major firms filed for bankruptcy in the 1980s. Military base closings and realignments will adversely affect the some 47,000 naval personnel and related civilian payrolls of more than $1 billion.

The economic growth of the metropolis enhanced the revenue base of local government. While the assessed valuation of real property was slightly above a billion dollars in 1964, the state of Florida mandated that the county tax assessor reassess property at market or "just" value. The 1965 reassessment resulted in an increase of about 150% in assessed value. By 1993, assessed taxable real property, at approximately two-thirds of "actual" value, was nearly $25 billion. Most of the ad valorem tax base had already shifted to the suburbs (about two-thirds by 1965).

The preconsolidation governments in the Jacksonville/Duval metropolis, especially the county and school district, relied heavily on property taxes. The former City of Jacksonville received three times more revenue from the "surplus" of its municipally owned electric utility than from property taxes (DeHoog & Swanson, 1988). By 1987, the consolidated government had substantially reduced its reliance on property taxes, from 62.6% to 29.7%. Intergovernmental revenues from the federal and state governments were increased from 30.6% to 38.2%. More significant is that the local tax burden from user charges shifted from 27.6% to 63.9%. Per capita revenues from their own sources increased nearly fourfold in current dollars, from a 1967 level of $265 to a 1987 level of $1,127. During the same time period, public officials shifted reliance away from property taxes, which increased only threefold, to less intrusive user charges, which increased more than seventeenfold. The cumulative effect of this shift stimulated one member of the Chamber of Commerce to note with dismay that city revenues had not been reduced, but the revenue package had been "nickel and diming" the average citizen of Jacksonville. The growth machine reduced property taxes to promote a very favorable business climate to attract economic growth and development.

Bahl, Mariez-Vasquez, and Sjoquist (1992) reported that fiscal disparities between the central city and suburbs changed little during the past two or three decades in metropolitan America. They explain intermetropolitan variations in fiscal disparities by

> differences in the communities' own resources, as proxied by per capita income, by differences in state and federal aid, and by how much of the jurisdiction services are demanded by nonresidents, as proxied by the employment to population ratio. In a post-aid world, differences in socioeconomic characteristics do not appear to play a significant role in explaining central city-suburban fiscal disparities. (p. 420)

Lineberry (1976) identified three competing criteria—equality, need, and demand—to justify urban service distributions.

Preconsolidated local governments imposed double taxation on inner-city residents, while residents of the suburbs received lower and unacceptable levels of public services. The expected economies from consolidation did not materialize, according to Benton and Gamble (1983), who found that the new government was spending as much as nearby Tampa, which had rejected merger at about the same time. They suggested that reformers may be "paying lip service" to efficiency, in fact having more interest in upgrading public services to residents in the unincorporated areas.

The preconsolidation county government chose not to maximize its tax base among suburbanites, who had a considerable higher socioeconomic status than those living in the City of Jacksonville. More important, inequities of service delivery persist, at least in the opinion of residents in many neighborhoods—older versus newer, minority versus majority, urban versus suburban, and upper versus lower income. The Northwest Quadrant, the core of the old city that contains most of the African American population, continues to experience high rates of poverty and crime, and it has poor stormwater drainage. Three black city council members were arrested for walking out of a city council meeting in protest; they later voted on the budget that they opposed for not addressing the problems of their districts. A recent study of neighborhood public services suggests that the city has not established criteria to measure and provide equitable service delivery, nor is the public sufficiently informed and involved in the decision making to modify the distributive patterns (Jacksonville Community Council Inc., 1994). The study proposed that the city establish standards and publish an annual "report card."

The Schizoid Metropolity

Expectations of dichotomous living in such proximate settlement patterns appear ominous to most reformers. They seem to perceive the metropolis as Greer (1962, p. 127) suggested, as "the snarl of politics, the multiplying suburbs, the slowly changing central city, becomes the given nature of things." The reformers cited civil unrest, social disparities, and governmental waste, fraud, and abuse as the basis for their vigorous drive for reunification of the polity. The Jacksonville metropoli had long been a place of dichotomous living by race and social class, as well as political participation, preferences, and sense of efficacy.

The merger of the governmental structures of the city and county was accomplished under the assumption that it would reunite the socioeconomic and political system. The initial and succeeding celebrations, at least for the first decade, testified that the objective had been achieved. The official report (City of Jacksonville, 1993, p. 2) on the 25th anniversary said that people have "good reason to be tremendously impressed and proud of what they have achieved for themselves through consolidation." More is required, however, than merging a fragmented government, as the metropolity remains schizoid.

The polity, long a traditional one-party Democratic stronghold of the South, began to change with the election of President Eisenhower in 1952. The electorate, divided by geographic residence (inner city versus suburbs), socioeconomic status, and race, differs markedly as to party registration, voter turnout, preferences, and representativeness.

The evolving two-party system reflects these differences, and political competition has increased. Less than 5% of the electorate was registered Republican prior to consolidation. By 1994, 30% of the electorate was Republican. Many of the Republicans were white, and whites composed three-quarters of the registered voters. Whites were about evenly split between being Democrats (52.8%) and Republican. African Americans, on the other hand, were overwhelmingly (93.5%) registered in the Democratic Party. Registration among suburbanites steadily increased to compose a substantial majority by the time of the consolidation election. While white registration declined in the city and increased in the suburbs, black registration increased, to two-fifths of the city electorate in the preconsolidation city, but then declined to one-fifth with consolidation. It remained at that level into the 1990s.

A systematic review of turnout during the decade prior to consolidation indicates a highly variable turnout record, depending on whether voters were choosing candidates for office in primaries or general elections and whether they were asked to respond to referenda issues such as bonds for improvement projects or charter reforms. The average turnout for city candidates was 52% in the first primary, 58% in the runoffs, and 27% in general elections. The dropoff in the general election reflects the ineffective challenges from the Republican Party in the mayoralty contests. The average turnout for referenda items was 43% during this same period, ranging from only 6% approving municipal bonds to expand the electric power system to 80% participating in the second unsuccessful annexation election preceding consolidation. Electoral turnout in the suburbs generally exceeds that in the inner city; however, in hotly contested races, turnouts are similar. The turnout of black voters exceeds that of whites when black candidates are on the ballot. For example, when Jesse Jackson ran in the 1984 presidential primary, black turnout was some 10% higher than white turnout.

Feiock (1992, p. 5) examined the effect of consolidation on electoral participation and found the average turnout to have been reduced by 17%. Consolidation diluted the black vote while promoting elite interests. He concluded that "Consolidation has had negative effects on citizen access and participation in government."

Electoral preferences among those living in the suburbs and those in the inner city differ dramatically. Duval County voters supported only Lyndon Johnson (1964) and Jimmy Carter (1976) as Democratic candidates. Support for independent candidates has developed mainly in the white suburbs, which gave George Wallace 36.7% of the vote and Ross Perot 13.4%. Black voters consistently have preferred Democratic candidates.

White suburbanites tend to support Republican and conservative Democratic candidates. Suburban voters elected a white Republican Congresswoman to replace long-term Democratic Congressman Bennett, and a minority district that follows Interstate Highway 95 from Orlando to Jacksonville and beyond elected a female black Democrat.

On referenda issues, city voters overwhelmingly (by a three to one margin) supported the annexation proposals of 1963 and 1964, whereas suburbanites opposed them by 56.2% in 1963 and 58.9% in 1964. Two-thirds of voters favored consolidation in 1967, irrespective of their residential location, socioeconomic status, or race. Similarly, in the 1990s there was broad support for tax caps and term limits for all local officials.

Districting 14 seats for the city council of the consolidated government provided representation for blacks and suburban Republicans. The size of the black population in Jacksonville—more than 165,000, the largest in Florida—has enabled African Americans to select their own representatives to state and local offices, and, with the reapportionment of 1992, to elect a black member of Congress. Black members of the city council compose their rough approximate share of the electorate, and one of those members served as president of the council. The suburban electorate has come to have considerable influence in the consolidated government, composing the majority of its electorate from the beginning. The majority of council members continue to be Democrats, but an increasing number of Republicans tend to be elected to the city council.

White suburban dominance was illustrated when the black council members walked out on a budget vote to protest what they believed to be unfair treatment of spending in their sector of town. The Republican president of the council, who was from the suburbs and later was elected to Congress, had them arrested and returned to the council chambers. Feiock (1992, p. 5) asserted that "In the short run, access of minorities can be guaranteed by drawing one or more minority districts, but in the long run minority representation is diluted."

The influence of political party affiliation will soon become less important. Voters overwhelmingly approved a "unified" ballot for local officials. The first four mayors were chosen in Democratic primaries; however, the incumbent was narrowly defeated in 1991 without the participation of suburban Republicans. Nor will incumbency be as important in the future, because the voters placed term limits on all locally elected officials.

Although the good intentions of the "centralists" were incomplete and generally remained so 25 years after the merger, the transformation of the power system became evident. The economy was expanding and diversifying, and it gave impetus to the perceived need for governmental reform. The old guard that controlled finance, railroads, timber, and land awaiting development also dominated local government, including the city and county levels, schools, and special independent authorities. Their ortho-dox conservatism of minimalist government and low property taxes was an impediment to those favoring growth, who wished to use the authority and public resources of local government to promote and nourish the agenda of land development. The reformer challengers of the 1960s not only wanted government to bring urban services to suburbanites but also wanted to expand economic development through such measures as sea and airport facilities and expressways that would make Jacksonville the "bold new city of the South."

Most studies of consolidation do not discuss the role of informal power systems in bringing about governmental reforms or the consequences of these reforms on the pattern of power and influence in the community. Erie, Kirlin, and Rabinovitz (1972, p. 28) state,

> Reform institutions do not, in the short run, alter the distribution of power in the metropolitan area. Unless previously inattentive groups are specifi-cally made aware of the differential advantages or deprivation inherent in institutional change, there is little penetration from these groups. The norm is vigorous competition among elites seeking relative advantage in the new institutions.

In the case of Jacksonville, Rosenbaum and Kammerer (1974, p. 50) explained the success of the consolidation vote as a matter of "power deflation," as people influential in the community became disillusioned with "inappropriate governmental responses" and "special interests" re-ceiving favors. They identified the reformers to be "top drawer" influen-tials of corporate, civic, and professional organizations.

Although reformers do not articulate the need for any redistribution of power, it is generally understood that there would be a rearrangement of power with a centralized, strong executive in control. The Jacksonville reformers not only reduced the influence of the Democratic Party but also, and more important, moved aside the dominant small economic clique that was unresponsive to demands for social and economic change. The prevailing power holders, politicians and economic elites, conservatively believed that city government should not be used to change the nature of race relations, nor should county government be used to plan and regulate land uses and provide public services in the growing unincorporated area.

Changing power structures generally is more difficult than simply restructuring authority or evicting politicians in a single election. The political struggles to establish the consolidated government involved evicting all elected officials of the city three months before the consolidation election. The reformers quickly filled the power vacuum by replacing the economic clique of landed and financial interests. The transformed power system also changed the prevailing ideology from orthodox conservativism and white supremacy to progressive conservatism and commitment to improving the provision of public goods and services. The power system brought forth a "growth machine" or development elite of landowners, developers, bankers, builders, and associated professionals including lawyers and engineers. Logan and Molotch (1988) emphasized the solidarity of the local business elite, who have a coherent strategy of control of the community. They use public institutions and personal influence to enhance their own self-interest.

Consolidation in Jacksonville did little to shift power from elites to the masses. Instead, it replaced the old guard political clique with the economically elite reformers, who set the community agenda that directs local government to facilitate the goals and objectives of the growth machine (Vogel & Swanson, 1989). Van Osdol (July, 1990) reported that power is in transition, switching from the old guard to modern sentries who blend new and old-line business leaders and a handful of politicians who work closely with the civic elite.

■ Conclusions

This report on how the consolidated government of Jacksonville provides regional governance has shown mixed results. Most of the socioeconomic and political profile characteristics did show and continue to show schisms that question the efficacy of centralized governance. Although consolidation provided the potential for regional politics of a metropolitan area, in Jacksonville and other consolidated cities it is confined to a single county. Although the Jacksonville MSA involves four counties, the prevailing sense of the metropolis has focused on Duval County. The single-tiered general consolidated government has incomplete control of the small towns, separately elected constitutional officers, and the quasi-independent authorities. The residents of the small towns say that they want their own county, and the constitutional officers are state officials who may appeal to the state cabinet for their budget requests. The independent authorities, which spend the largest share of public funds, serve as the engine of growth for Jacksonville.

The most integrative force in Jacksonville has come from the informal power system, which has been significantly transformed from a tight, consensual, inner clique that used oligarchic control to impose its orthodox political ideology of a minimum government with low taxes. The new consensual elite has become a "growth machine" with an operative ideology of progressive conservatism toward an expansive use of governmental authority and public resources to facilitate growth. Although this growth machine promotes regional perspectives, it still lacks consensus about development of the downtown area and locating business in the largely black Northwest Quadrant of town.

Although few political participants would abandon the merger, positive citizen perceptions of local leadership have fallen perceptively, from 50% in 1985 to 29% in 1993 (Jacksonville Community Council Inc., 1993). This decline reflects the rise of urban challengers to protect environmentally sensitive lands, neighborhoods, and the black community (Swanson, 1992). Voters have expressed anger over their sense of paying more and getting less, the number of city employees being reduced, residents in the old city being asked to pay for garbage collection for the first time, and the belief that public services are inequitably delivered (Jacksonville Community Council Inc., 1994). In the early 1990s, voters evicted one mayor and stimulated his successor not to seek reelection. Voters imposed term limits on all elected officials, tax caps of 3% above the rate of population growth on property assessments, and the "unitary" ballot, which allows voters to cast ballots for candidates of any party irrespective of party registration. Unsuccessful efforts to "reform the reform" continue, with attempts to include the small towns (three beach towns have petitioned to create a new "Ocean" county) and independent authorities in the consolidated government and to reduce the size of the city council.

Jacksonville's domination of the regional economy of northeast Florida, known as the First Coast, has been enhanced by efforts of the Jacksonville Chamber of Commerce to promote its economically oriented agenda. The state mandated that all localities prepare comprehensive land-use plans that are consistent with the state and regional plans. Failure to manage growth and provide public infrastructure to meet population growth could result in a state-ordered moratorium on growth.

Governing the schizoid metropoly, even with a centralized government, is extremely difficult. Public officials and civic leaders must process the claims of different interests and produce definitive decisions on conflicting and uncertain positions. They are faced with providing a complex set of costly services with shrinking funds provided by a resisting electorate. The problem of equity resulting from "double taxation" of city residents and the "free rider" phenomenon of suburbanites consuming city services without payment largely has been resolved. Suburbanites have not escaped the costs of supporting the growing underclass in the city, which pays less in taxes than it consumes in public services. To do so, they will have to move to adjoining counties that are not part of the Jacksonville consolidated government.

Perhaps more difficult are the problems associated with building consensus to plan the future direction of the community. Both public officials and economic leaders are faced with seeking ways to relate governmental redevelopment plans for central business district and slum clearance, while accommodating the marketplace proliferation of shopping malls, office complexes, and industrial activities in areas that require heavy public investments in infrastructure.

To build a viable metropoly should take into account what Lehman (1992, p. 59) identified as

> three pivotal anomalies of modern democratic life: (1) why enhancing effective government capacities in one sector can undermine overall effectiveness and even foster inefficient participation and perhaps declining legitimacy; (2) how polities can be simultaneously racked by apparently contradictory inefficiencies of extensive public apathy and widespread moralistic participation; and (3) why high political legitimacy actually goes hand in hand with pervasive lack of confidence in government leaders.

Lehman noted that a polity does not completely succeed or fail; it is a matter of the degree to which it is viable. He cited the centralization of government and potential overemphasis on economic development at the expense of quality of life concerns of the populace. It is important to design a polity in which political outcomes can be achieved in the face of apparently contradictory inefficiencies of extensive public apathy and widespread moralistic participation.

It will take more than the structural reform of government to cope with festering social and racial disparities. American federalism was designed for a plurality among institutions of governance, but it will take Ostrom's (1991) *res publica* of public opinion, civic knowledge, and a culture of inquiry to make it work. This involves accepting, even preferring, a dynamic, polycentric set of policy decision centers that serve as complementarities in discovering the regional or whole community.

REFERENCES

Advisory Commission on Intergovernmental Relations. (1965). *Metropolitan social and economic disparities.* Washington, DC: Government Printing Office.

Advisory Commission on Intergovernmental Relations. (1974). *Governmental functions and processes.* Washington, DC: Government Printing Office.

Anderson, W. (1925). *American government.* New York: Henry Holt and Co.

Bahl, R., Mariez-Vasquez, J., & Sjoquist, D. L. (1992). Central city-suburban fiscal disparities. *Public Finance Quarterly, 20*(4), 420.

Benton, J. E., & Gamble, D. (1983). City/county consolidation and economies of scale. *Social Science Quarterly, 65*(1), 190-198.

Brown, L. D. (1978). Mayors and models. *American Politics and Public Policy.* Boston: MIT Press.

City of Jacksonville. (1992). *Insight.* Jacksonville: Author.

City of Jacksonville. (1993). *Consolidated Jacksonville.* Jacksonville: Author.

DeHoog, R., & Swanson, B. (1988). Tax and spending effects of municipal enterprises. *Public Budgeting & Finance.*

Erie, S. P., Kirlin, J. J., & Rabinovitz, F. F. (1972). Can something be done? In L. Wingo (Ed.), *Reform of metropolitan governments.* Baltimore: The Johns Hopkins University Press.

Farley, R., & Taeuber, A. F. (1974). Racial segregation in the public schools. *American Journal of Sociology, 4*(79).

Feiock, R. (1992). The impacts of city/county consolidation in Florida. *Governing Florida, 2*(2), 3-5.

Greer, S. (1962). *Governing the metropolis.* New York: John Wiley & Sons.

Greer, S. (1963). *Metropolitics: A study of political culture.* New York: John Wiley & Sons.

Jacksonville Community Council Inc. (1993). *Live in Jacksonville.* Jacksonville: Author.

Jacksonville Community Council Inc. (1994). *Jacksonville public services.* Jacksonville: Author.

Lehman, E. W. (1992). *The viable polity.* Philadelphia: Temple University Press.

Lineberry, R. L. (1976). Who is getting what? *Public Management, 58*(8), 13-18.

Local Government Study Commission of Duval County. (1966). *The blueprint for improvement.* Jacksonville: Author.

Logan, J. R., & Molotch, H. L. (1987). *Urban fortunes.* Berkeley: University of California Press.

Martin, R. (1968). *Consolidation.* Jacksonville: Crawford.

Miller, D. C. (1975). *Leadership and power in the Bos-Wash megalopolis.* New York: John Wiley & Sons.

Morgan, D. R., & Pelissero, J. P. (1980). Urban policy: Does political structure matter? *American Political Science Review, 74*(4), 1005.

National Advisory Commission on Civil Disorders. (1968). *The Kerner report.* Washington, DC: Government Printing Office.

Ostrom, V. (1991). *The meaning of American federalism.* San Francisco: Institute of Contemporary Studies.

Rosenbaum, W. A., & Kammerer, G. M. (1974). *Against long odds.* Beverly Hills, CA: Sage.

Stien, M. R. (1960). *The eclipse of community.* New York: Harper.

Swanson, B. E. (1970). *Concern for community in urban America.* New York: Odyssey.

Swanson, B. E. (1992). *Early detection of emergent patterns of social and political challengers.* Paper presented to the Florida Political Science Association.

Swanson, B. E., & Swanson, E. P. (1979). *Discovering the community.* New York: Irvington.

Taeuber, K. E., & Taeuber, A. F. (1965). *Negroes in cities.* Chicago: Aldine.

Van Osdol, P. (1990). Jacksonville's ten most powerful. *Florida Times-Union.*

Vogel, R. K., & Swanson, B. E. (1989). The growth machine versus the antigrowth coalition. *Urban Affairs Quarterly, 25*(1), 63-85.

11

Portland:
The Metropolitan
Umbrella

ARTHUR C. NELSON

■ Introduction

The Metropolitan Service District, which encompasses the contiguously urbanized areas of the Portland, Oregon, Metropolitan Statistical Area,[1] is a multipurpose regionally elected governing body that is at once pathbreaking as a mode of regional governance yet benign in its functions. "Metro" provides several services that cut across traditional municipal and county boundaries of 24 cities and three counties. These services include a regional zoo, regional solid waste disposal, and regional tourism development. Metro also is responsible for coordinating growth management, land use, and transportation planning within the region, and it has the authority—not yet exercised—to assume management of the region's transit operations.

Metro is the nation's only directly elected regional governing body. The Metro Council is composed of seven nonpartisan members elected from seven districts of roughly comparable population (about 200,000 each) for 4-year terms. The Council is responsible for Metro's policy formulation, legislation, and budgeting. Daily management of Metro is the responsibility of an executive officer who is elected at large.

Metro was authorized by the Oregon legislature in 1977 and approved by voters of Clackamas, Multnomah, and Washington Counties in 1978. It replaced the Columbia River Association of Governments (CRAG), which had served as the Portland-Vancouver metropolitan area's regional council of governments (COG) since 1966. CRAG itself replaced the Portland-Vancouver Metropolitan Transportation Study (PVMTS), formed in 1959. In 1992, voters in the region approved a home-rule charter for the agency and, among other things, changed the Metropolitan Service District's name to simply "Metro."

The revenue base for Metro is composed of solid waste tipping fees and zoo admission fees including an excise tax thereon, regional property taxes dedicated to operating the zoo and retiring revenue bonds that finance the Oregon Convention Center, and a Multnomah County transient occupancy tax that is used to support Convention Center operations located in that county. Subject to regional voter approval not yet tested, Metro's revenue base can include a sales tax and an income tax.

In many respects, Metro is a model of metropolitan governance that may be replicated by other metropolitan areas over the next generation. Although some may contend that Metro's rise is attributable to metropolitan Portland's unique political-economic climate and the governance concept it represents is not generalizable, this is a mistaken view. Metro's rise is attributable more to finding ways to solve problems common to all regions than to anything peculiar about metropolitan Portland. The only peculiarity about metropolitan Portland is that its particular governance structure was the first of its kind in the nation, but it certainly is not the last. After reviewing Metro's political-economic context and discussing the metropolitan governance context, this chapter presents the features of Metro that are generalizable.

■ Political-Economic Context

The Portland metropolitan area is situated at the confluence of the Willamette River and the Columbia River. The Portland-Vancouver consolidated metropolitan statistical area (CMSA) is composed of five counties. The central county within which Portland is predominantly situated is Multnomah. The suburban counties are Clackamas, Washington, and Yamhill on the Oregon side of the Columbia River, and Clark County (the central county of Vancouver) on the Washington side. Yamhill county is the sole exurban county within the MSA; however, because of very large land areas, much of the suburban counties and the eastern quarter of the central county also can be considered exurban, except that populations in the exurban portions of those counties are quite small. As seen in Figure 11.1, Metro serves an area smaller than both the Portland-Vancouver CMSA and the Portland MSA; however, it encompasses nearly the entire

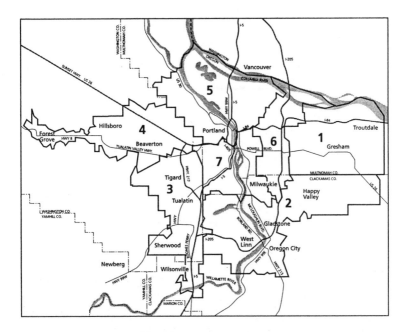

Figure 11.1. Metro's Boundaries and Councilor Districts
SOURCE: Metropolitan Service District, copyright 1994.

population of the three Oregon counties within which it operates. Population trends for these three counties since 1970 are reported in Table 11.1.

Portland began as a port for shipping timber and agricultural products and importing finished goods, in the late 1840s. It was incorporated in 1851 and named Portland because Francis W. Pettygrove of Portland, Maine, won a coin toss used to determine the port's name over Asa L. Lovejoy of Boston, Massachusetts. Between 1883 and 1910, when Portland was connected by rail to Chicago and other points east, Portland was second only to San Francisco as the West Coast's largest city. In 1933, the "Tillamook Burn" destroyed much of the timberland accessible to Portland and the city lost much of its economic base. During World War II, more than 100,000 people moved into the city to build ships for the Pacific theater, and many tens of thousands stayed after the war. In the 1970s and 1980s, the city undertook several major downtown revitalization efforts. During the same time, Oregon adopted sweeping land use plan-

TABLE 11.1 Population Trends, 1970-1990

Characteristic	Portland Central City	Portland MSA
1970 population	382,619	878,676
1980 population	366,383	1,050,418
1990 population	437,319	1,174,291
Change 1970-1990	54,700	295,615
Percentage change 1970-1990	14.30	33.63
Regional growth share (percentage)	18.50	100.00

SOURCE: *State and Metropolitan Area Data Book 1991*, Bureau of the Census, 1991.

ning laws that preserved farmland and forest land around Portland for resource uses and directed urban development into Portland and its nearby suburbs.

Although it began as a center for shipping raw products and importing finished products, Portland has become among the nation's most diversified metropolitan economies, second only to Chicago in indices of diversity. More than 2,800 manufacturing plants are located in the Portland metropolitan area. About one-eighth of the workforce is engaged in manufacturing. The leading manufacturing industry is metal processing, followed by electrical equipment, food products, lumber and wood products, paper, and transportation equipment. The Port of Portland, a statutorily authorized economic development district, handles more lumber and agricultural cargo than any other West Coast port. Since the 1960s, however, Portland has lost manufacturing employment while the metropolitan area has gained, a trend that is reported in Table 11.2. This is consistent with post-World War II trends wherein manufacturing has become suburbanized. On the other hand, the total civilian labor force has grown substantially in both Portland and the region, as seen in Table 11.3.

The tables and discussion clearly show that Portland's share of regional development has been falling gradually with respect to its suburbs since the 1960s. This is not unexpected, as it follows national trends. On balance, however, metropolitan Portland remains a monocentric urban region. As many metropolitan areas became polycentric during the 1980s, primarily as a result of the construction of more suburban office and retail space than exists in downtown, Portland's downtown has remained the region's dominant center. By 1990, for example, downtown Portland accounted for about 60% of the region's office space and half of its upscale

TABLE 11.2 Manufacturing Employment, 1963-1987

Characteristic	Portland Central City	Portland MSA
1963 employment	35,646	56,552
1987 employment	32,700	89,100
Change 1963-1987	−2,946	32,548
Percentage change 1963-1987	−8.26	57.55

SOURCE: *Census of Manufactures*, Bureau of the Census, 1963a and 1987a.

TABLE 11.3 Total Employment, 1963-1987

Characteristic	Portland Central City	Portland MSA
1963 employment	146,385	211,088
1987 employment	202,244	459,800
Change 1963-1987	55,859	248,712
Percentage change 1963-1987	38.16	117.82

SOURCE: *County Business Patterns*, Bureau of the Census, 1963b and 1987b.

retail space. Although two "edge cities" have formed southeast (Clackamas Towne Center) and southwest (Washington Square) of downtown, they are not being designed to exceed the downtown office and retail space, as has happened in other metropolitan areas.[2] Eventually, total office and upscale retail space in the suburbs will exceed that of downtown, but downtown will remain the region's focal point, having the largest concentration of office and retail workers, postsecondary education students and teachers, multifamily housing units, and hotel rooms and restaurant seats.

■ The Metropolitan Governance Context

Metro is an integral part of the metropolitan governance context and, in turn, cities, counties, and special service districts look to Metro for coordination of planning but little else. To understand this relation, two pieces of background information are needed. The first is a brief historical review of the rise of a metropolitan governance ethos; the second is a discussion of the interplay among the mosaic of governance structures, both formal and informal, that characterize the region. The section concludes by portraying Metro as a kind of "umbrella" for the metropolitan region.

Rise of a Metropolitan Governance Ethos

Between its incorporation in 1856 and 1906, the City of Portland grew through annexations and mergers to remain the only urban government in its region. Portland owed its expansion success to a state legislature that, among other things, merged Portland with its early suburban cities of Albina, East Portland, Sellwood, St. Johns, and Linnton. In 1906, Oregon voters approved a constitutional amendment allowing home rule to cities whose voters adopted home rule charters. This effectively stopped annexations of larger cities such as Portland because, in order for territory to be annexed, approval had to be obtained from at least half of the voters of the territory requesting annexation (City Club of Portland, 1986). Later, a "triple" majority could only affect annexation: a majority of the voters representing a majority of the land area and a majority of the total property value. As suburban areas grew and needed governance structures to manage facilities and services, the trend was for voters to incorporate into independent cities. Voters also possessed the power to create limited or special service governments to manage specific services within unique boundaries and with a dedicated property tax base. By 1956, some 50 years after the home rule constitutional amendment, the tri-county area (Clackamas, Multnomah, and Washington counties) was composed of 176 separate units of government. Statewide, there were 218 special districts for fire, lighting, parks, sewers, water, and zoning, making Oregon, with only 1% of the nation's population, the state with the seventh largest number of such districts (Abbott & Abbott, 1991, p. 7).

In the 1950s, elected officials including municipal and county officers and state legislators had become concerned about the efficiency and effectiveness of the delivery of services. Many special districts were too small to provide services economically, so service was either shoddy or expensive, or sometimes both—raising the ire of taxpayers within those districts. To help make the delivery of service more efficient and less costly to the taxpayer, the Oregon legislature enabled counties to absorb special districts into countywide single-purpose special districts managed by the board of county commissioners. By 1976, 20 years later, the number of special districts had been reduced to 137, although nine new cities and five new regional entities were formed in the meantime (Abbott & Abbott, 1991, p. 585).

During the 1970s and early 1980s, four highly significant changes in the way in which services were delivered to suburban areas occurred. The first was the formation of a region sewerage agency in 1970, serving mostly suburban Washington County west of Portland. The second was a fiscal crisis facing Multnomah County's management of unincorporated suburban eastern Multnomah County in the early 1980s. The third was the formation of the Metropolitan Service District in 1970, which was then called MSD and possessed limited responsibilities, with even more limited funding. Fourth, the Columbia Region Association of Governments was abolished and its functions transferred to a reconstituted MSD, the name of which was changed in 1992 to "Metro."

The Once and Future Metropolitan Service District. Despite legislative efforts in the 1950s to make government more efficient and less costly to taxpayers, there was growing public concern in the metropolitan area about the proliferation of uncoordinated special districts and newly formed cities competing for growth. In 1960, the League of Women Voters published *A Tale of Three Counties,* which voiced concerns about uncoordinated services and wasteful spending. The publication pointed to the need for new governance structures that were based on efficiency and accountability. The Portland business community, through its Chamber of Commerce, joined in the call to rethink government structures to make the delivery of services more efficient and responsive. The legislature in 1961 created the Portland Metropolitan Study Commission (PMSC), and 38 people were selected to staff the commission (one by each of the state legislators representing the tri-county area). The commission was composed of a wide range of prominent elected officials, business leaders, and citizen activists. The PMSC operated from 1963 to 1971 and recommended, among other things, a regional transportation authority (Tri-Met) and a metropolitan service district. Both were enabled by the legislature in 1969 and adopted by metropolitan voters in May of 1970. Oddly, although metropolitan voters gave Tri-Met the authority to impose a payroll tax to fund regional transit in the November, 1970, general election, voters denied MSD a property tax base. The MSD governing board was then composed of ten members, one appointed by the Portland City Council, two each appointed by the three county commission boards, and one each appointed collectively by the cities within each county (Abbott & Abbott, 1991, pp. 7-11, 20-25).

Although the MSD was given extensive powers by the legislature in 1969, and by powers approved by metropolitan voters in 1970, for several years MSD conducted only regional solid waste planning. In part, the lack of a tax base kept it from taking on additional functions. Small loans from the state Department of Environmental Quality and a small tax on the disposal of used tires were MSD's principle sources of funds (Abbott & Abbott, 1991, pp. 15-16).

In 1976, the City of Portland faced a crossroads in operating its Washington Park Zoo, which drew more nonmetropolitan residents than Portland residents. Needing an infusion of capital, Portland faced three major choices: let the zoo become less well managed and eventually close it, raise fees in part by assessing non-Portland residents more than Portland residents, or transfer the zoo to a regional agency with a regional mandate to run it. The MSD agreed to take over the zoo subject to metropolitan voters approving a property tax base for it, which the voters did (Abbott & Abbott, 1991, p. 16). Thus, by the end of 1976, the MSD provided regional garbage disposal planning and ran the region's zoo. From this humble beginning, the MSD would be reconstituted in 1978 as one of the nation's most ambitious metropolitan governance experiments.

A Unified Sewerage Agency Serving Western Suburbs. During the 1960s, Washington County was rapidly growing into a series of incorporated and unincorporated suburbs without coordination of water, sewer, or road facilities. It was the crisis in the delivery of sewer services that led a coalition of taxpayers and developers to push for the merger of 23 sewer districts into a "Unified Sewerage Agency" (USA) that was passed by voters in the affected area in 1970. Among the reasons cited in support of the USA formation were: that existing sewer facilities could not accommodate growth in areas where growth pressures were the greatest; costs of running small sewer districts were rising rapidly, yet service was deteriorating; lack of coordination in sewer facility planning stifled economic growth; and developers were frustrated in having to work with many separate sewer districts, each with its own technical requirements and review procedures. In retrospect, the merger of small sewer agencies into the regional USA clearly was intended to facilitate growth.

The USA does not affect much of Clackamas or Multnomah Counties, where more than a dozen sewer agencies operate. Some have suggested that the USA's more efficient processing of sewer permits—which are linked to land use permits—and delivery of sewer services coordinated among all affected cities is the single most important reason why Washington County has emerged as the region's largest suburbanized area and the state's wealthiest county, even though suburban Clackamas and suburban eastern Multnomah Counties were of similar population size, growth rates, developable land areas, income, and economic development resources in 1970 (Nelson, 1987; Nelson & Knaap, 1987). Perhaps the efficiency with which USA manages Washington County's sewer system was a factor in its voters supporting the reconstitution of MSD in 1978.

Fiscal Crisis in Eastern Multnomah County Suburbs. In 1983, Multnomah County faced large budget deficits and a growing need for services in the suburbanizing but unincorporated areas east of Portland. To solve the crisis, the Multnomah County Board of Commissioners adopted "Resolution A," which directed the county to negotiate intergovernmental agreements with Portland and suburban Gresham to provide urban services. The county also facilitated the annexation of much of the unincorporated area into Portland or Gresham. To facilitate efficient delivery of sewer services, the county engineered the formation of a joint sewer construction program involving the county, Portland, Gresham, and Troutdale (City Club of Portland, 1986, p. 587). The fiscal and managerial benefits of merging services have been seen in Multnomah County, and it is perhaps in part for this reason that Multnomah County voters supported the reconstitution of MSD in 1978.

The Arrival of "Metro." Although CRAG was formed in 1966 to coordinate regional planning, it was viewed as largely ineffective for three reasons. First, like most COGs in operation then and today, CRAG's board was composed of representatives of municipalities and counties whose primary interests were their home constituencies. Municipalities and counties were in direct competition for growth, and regional plans threatened their ability to attract desirable growth. Second, CRAG depended on federal funds and voluntary membership of local governments, along with their voluntary dues. Third, with Portland being the largest jurisdiction and possessing the most professional government staff in the state, smaller cities and understaffed counties believed themselves to be at a disadvantage in analysis and policy making.

The second problem was solved in 1973, when the Oregon legislature made CRAG membership mandatory and required members to pay dues on a per capita population basis. Although this gave CRAG reasonably solid funding for its operations, the other two problems left it mostly ineffective in influencing long-term regional planning solutions. One of CRAG's limitations was a lack of constituency, leading the *Oregonian* newspaper to call CRAG a "stranger to the people it serves" (Abbott &

Abbott, 1991, pp. 17-19). Despite its shortcomings, CRAG became the nation's first regional body granted legislative authority to require cities and counties to conform their land use plans to regional standards (Leonard, 1991, pp. 98-99).

In 1975, another study commission was formed to consider, among other things, how best to manage regional problems while retaining local control. Through a grant from the National Academy for Public Administration matched by local businesses, local governments, CRAG, Portland State University, and the Metropolitan Boundary Commission (formed in 1969 to oversee annexations and special district formation to avoid inefficient duplication of government services), the "Tri-County Local Government Commission" was formed. In 1976, it recommended that functions of CRAG be transferred to the MSD in part because the MSD already had a regional electoral mandate and had proven itself efficient in its operations, and in part because of irreparable animosity toward CRAG. The MSD's board would be replaced by an elected body composed of councilors elected from districts and an executive director elected at large. These recommendations were designed in part to give the MSD a political constituency through an elected board that would be held accountable to the voters for their behavior (Abbott & Abbott, 1991, pp. 20-25).

To prevent the MSD from being viewed as threatening to incumbent legislators from the region, a large number of councilors—15—was recommended; this would result in the MSD districts being smaller than state senate districts. The Commission recommended that district boundaries conform to historic or traditional communities rather than existing political boundaries. This had the additional effect of binding smaller cities and unincorporated areas within the same geography into districts of their own, while also ensuring larger cities and unincorporated areas of representation on the board (Abbott & Abbott, 1991, pp. 20-25). There was yet a third effect of this approach. Although Portland would have the largest number of councilors, it would hold a minority of seats because Portland's population was barely more than a third of the regional population; thus, whereas Portland had dominated staff-level involvement with CRAG, its influence would be considerably weaker within a regionally elected governing body. Suburban representation would be stronger.

In 1977, the legislature modified the Commission's recommendations and referred the question of whether to "Reorganize Metropolitan Service District, Abolish CRAG" to metropolitan voters in May, 1978. The reconstituted "Metro" won 55% of the vote. A majority of voters in Multnomah and Washington Counties supported Metro, while it received about 2,000 votes less than a majority in Clackamas County. It is this writer's observation that on the whole, voters in Multnomah and Washington Counties have had favorable experiences with the merger of smaller special-purpose districts into regional districts, whereas Clackamas County had no such political experience (see also Abbott & Abbott, 1991, pp. 24-25).

In November, 1978, the nation's first elected regional governing body was formed, and the newly constituted Metro board—composed of 12 councilors and an executive officer—became effective on January 1, 1979. Later in 1979, the Oregon Land Conservation and Development Commission approved the metropolitan urban growth boundary (UGB) designed by Metro (Abbott & Abbott, 1991, pp. 26-27), creating the nation's most complex regional urban containment policy. For its part, the UGB reflects consensus among more than 20 cities and three counties on how to allocate growth among them.

A Mosaic of Governance Structures

It is important not to gather the impression that Metro is an all-powerful regional government. In fact, Metro operates within a sea of other governance structures, and its budget is piddling by comparison to many other governmental units. Excluding school districts, there remain more than 100 taxing jurisdictions within the metropolitan area, more than 80 of which are single-purpose special districts. There are also four counties and more than 20 cities that perform multiple services, some of which overlap Metro's, in addition to the Vancouver MSA. Moreover, there are two formidable regional entities over which Metro has little or no control. It is important to appreciate its relationship with these groups of jurisdictions to understand how Metro enjoys some degree of political success.

Relations With Special Districts. One of the keys to Metro's success is that it limits its jurisdiction to only those functions for which there is consensus on a regional role. The zoo, garbage disposal, and coordination of regional development patterns are legitimate functions of a regional agency. In such cases, economists would consider these functions "public goods" and therefore appropriate for regional agencies to manage.[3] In contrast, the efficient provision of water, sewers, police and fire protection, libraries, and similar services depends greatly on population density, growth patterns, physical terrain, technology, and the extent to which users (usually also taxpayers) are willing to pay for higher or lower levels

of services. Many such services are most efficiently provided in arrangements that may appear haphazard because they cross city and county
boundaries, and even cross themselves. In fact, when properly designed,
single-purpose special districts are likely to be more efficient than if
management was through a centralized governance structure.

Metro understands this, perhaps only intuitively, and is not interested
in acquiring these responsibilities. This gives special districts little or no
reason to fear Metro's presence. On the other hand, Metro provides these
special districts with the forum within which special districts may coordinate planning and investment activities among them, and Metro can
serve as an umbrella agency in using its professional staff to solicit federal
or state funds to help support certain activities of special districts. Finally,
through its databases and analytical capacities, Metro can assist small
special districts in anticipating future needs.

Relations With Other Multipurpose Jurisdictions. More interesting is the
relationship between Metro and the multipurpose jurisdictions, which are
the three urban counties and the more than 20 cities within Metro's
boundaries. In some respects, Metro's services overlap these jurisdictions,
especially in the area of planning. Each county and city prepares and
implements its own land use and development plan. For its part, Metro
coordinates the preparation of those plans at the regional scale. Under
Oregon planning law, local governments must coordinate plans with their
neighbors. Regional agencies such as Metro have special statutory
authorization to review plans and offer recommendations to ensure that
plans are properly coordinated. Local governments that fail to cooperate
risk important legal sanctions. As a practical matter, the cities and counties
within the region work in good faith to coordinate their plans more out of
the fear of coordinated attacks by other jurisdictions than for altruistic
reasons or the fear of state sanctions.

An example is illustrative. Because of the success of a light rail line
—called MAX—connecting downtown Portland east to Gresham, Metro
secured regional voter approval and federal matching funds to build a west
line from downtown Portland through several Washington County suburbs, ending in Hillsboro some 20 miles away. As one may imagine, the
selection of the route should have been hotly contested among competing
communities. Instead, a deliberate planning process led to a routing that
best achieved development goals of competing jurisdictions. That process
also resulted in the design of long-range plans to extend light rail lines to
most remaining suburbs throughout Washington and Clackamas Counties.

Another example also is illustrative. Metro essentially has the final say (before state courts) on whether and when the UGB should be expanded to allow more urban fringe development. In theory, the UGB should not be expanded until it is first substantially filled in. Development patterns, however, have been such that demand for development in the southwestern part of the UGB is considerably greater than in some eastern sections, prompting some southwestern jurisdictions to lobby for UGB expansion. Generally speaking, the Metro board has opposed these moves, opting for the view that one purpose of the UGB policy is to redirect development away from areas that are being built or are already built to other areas where development capacity remains. This policy is having the effect of redistributing development pressures so that, eventually, all jurisdictions enjoy development over time. Indeed, because the existing UGB is rapidly running out of space, Metro is engaged in a comprehensive, multijurisdictional planning process to decide how to accommodate regional development between 2000 and 2040.

Relations With Other Major Regional Agencies. Metro is not the region's only regional government. Tri-Met, which runs the metropolitan area's transit system, and the Port of Portland, which operates the region's ports, largest industrial parks, and airports, are in some ways more influential.

The 1970 referendum creating the MSD, later called Metro, authorized Metro to absorb Tri-Met. Although this authority remains, it has not been exercised. One reason is simply that Tri-Met is considered to be reasonably well run and, as the sentiment goes, if something is not broken, it should not be fixed. More to the point, Tri-Met runs buses and trains, sets routes, hires and fires people, and manages a budget. These are responsibilities that are not normally well suited to multipurpose agencies, especially those without a history of managing complex systems. The second reason is that although Tri-Met runs transit, Metro does the long-range transportation system planning for Tri-Met. For example, it is the agency sponsoring light rail planning, design, and funding. In this respect, each agency does what it does best within a coordinated regional context.

A more distant relationship exists between Metro and the Port of Portland. The Port of Portland was created by legislative act in 1891. Its original tax base came from Multnomah County, and its principal responsibility was dredging the Willamette and Columbia rivers to facilitate ocean shipping.[4] In 1974, metropolitan voters extended the Port's tax base to include Clackamas and Washington Counties. The Port manages the nation's busiest inland river port, one of the nation's largest industrial parks, the region's international airport, and several commuter airports. Although the Port's land use plans must be coordinated through Metro, that is nearly as far as the relationship goes.

Relations With Metropolitan Cities and Counties Outside Metro Boundaries. Metro's jurisdiction does not include the entire metropolitan area as defined by the U.S. Census, nor does its jurisdiction cross state lines to include Vancouver, Washington. Because Metro is an *urban*-oriented regional government and because Oregon law prevents urban development from occurring on rural or agricultural lands, Metro's boundaries were designed consciously to serve only the present and future urban areas, and no more. Conceivably, over time and as the UGB may be expanded into agricultural districts, Metro's boundaries may also expand.

Although the exclusion of these other local government units from Metro makes it less "regional" in scope, there are other mechanisms in place to coordinate the interests of these other jurisdictions with Metro. For example, although Yamhill County is excluded entirely from Metro's jurisdiction and rural portions of the other three Portland MSA counties are also excluded, the municipal and county governments are nevertheless brought together for regional planning purposes through the Joint Policy Advisory Committee on Transportation (JPACT)—staffed by Metro—to coordinate federally mandated transportation planning. To receive federal transportation funds, every Census-defined metropolitan area in the United States must create a metropolitan planning organization (MPO) that provides coordinated transportation planning throughout the MSA. JPACT serves this function for the Portland MSA. JPACT also coordinates transportation planning with the Vancouver MPO, which is necessary for MPOs within consolidated MSAs. Transportation is recognized as an important contributor to land use patterns and environmental quality by the federal Intergovernmental Surface Transportation Efficiency Act (ISTEA) and the federal Clean Air Act Amendments, both passed by Congress in 1991. Although Metro's jurisdiction is limited to its boundaries, coordinated regional transportation and land use planning is conducted through JPACT.

Still, in terms of land use planning, the MSD covers considerably less territory than CRAG. The MSD's planning powers are limited to its boundaries, whereas CRAG's former boundaries encompassed all of Clackamas, Multnomah, and Washington Counties. The MSD thus focuses on coordinating planning within the metropolitan area and not truly integrating planning done throughout the entire Census-defined metropolitan area (Leonard, 1991, p. 100).

The Metropolitan Umbrella Role

Although political scientists may call metropolitan Portland's governance structure a "two-tiered" system, this is not an entirely accurate portrayal. Two-tiered governance structures are usually conceived of as having a regional entity charged with managing all services of a regional nature, with local entities managing strictly local services. Seattle's Metro, for instance, was formed in 1958 to rid Lake Washington of pollution, and it has since become the region's provider both of sanitary sewerage collection and treatment and of mass transit. The Minneapolis-St. Paul Twin Cities Metropolitan Council not only provides coordinated planning, but also supervises regional transit, airport services, and waste control; it also has become a de facto cable television coordination body, housing authority, and park acquisition authority.[5] Metropolitan Dade (Metro Dade) County, Florida, provides multiple services throughout the county with two budgets: one for countywide operations, financed from countywide taxes, and another for services to unincorporated areas, financed from special taxes on such areas. In Seattle and the Twin Cities, all other public services are provided by local government, whereas in Dade County, municipalities provide all other public services.

Metro does not fit any of these molds. One could argue that its management of the zoo and regional landfills (but not collection, rate setting, routing, etc.) qualifies it as a regional multipurpose service agency, but this is stretching the concept. More to the point, it does not provide regional water, sewer, police, airport services, parks (other than the zoo), public housing, cable television, or any related "service." It does not run the transit system, although it provides transportation planning services.

If Metro really is not a two-tiered governance structure, what is it? Would it be a three-tiered system, in which coordination is performed at the top tier, single purposes that are regional in scope are performed within the second tier, and local governments perform all other services? This cannot be, because none of the other conceivable second-tier regional agencies provides more than one service; they are merely large special districts. Obviously, Metro is not a first-tier government.

The argument is offered that Portland's Metro really is an "umbrella" under which a variety of policy-related functions are performed, but all services in the usual sense are provided by local or regional single-purpose governments. The elected board of Metro has no power to control the behavior of local governments; its primary authority other than setting zoo admission fees and garbage tipping fees is coordinating planning. In this respect, its greatest single influence is in transportation planning because, as the region's staff to JPACT, it influences the allocation of federal transportation funds. In this respect, JPACT is no different from any of the other some 400 MPOs found across the nation.

The conclusion drawn is that Metro functions as an umbrella. An umbrella is, generally, a device that shelters persons from the elements. Perhaps Portland's celebrated rainfall gives special meaning to the word. As an umbrella, Metro brings together local and regional interests under one roof, from coordinating a variety of issues ranging from fair share housing allocations to UGB adjustments to appropriating of transportation funds to simply creating consensus for new regional initiatives. In recent years, such initiatives have included comprehensive light rail planning, efforts to seek voter approval of general obligation bonds to acquire the development rights from farmland outside the UGB to preserve such land for farming in perpetuity, and selecting the site (with considerable political aplomb) for the region's $65 million Oregon Convention Center.[6]

■ More Metropolitan Umbrellas?

Portland's Metro is an experiment in governance. Its beginnings as the MSD were humble, providing only solid waste disposal planning services and eventually running a regional zoo. After having mastered these limited functions, Metro became the vehicle through which coordinated regional planning would be legitimized, principally through the election of a governing body. Some could argue that Metro's rise is attributable to something unique about metropolitan Portland's political economy, yet, as this chapter has demonstrated, the only thing unique about Metro was the interest of vested groups in providing more effective services benefiting all vested interests, where economies of scale were found to be regional in scope in specific areas. Otherwise, Metro was intentionally designed to have limited scope or authority over local or special-purpose interests. Indeed, one could argue that the idea of a regionally elected body in effect advances the interests of vested groups rather than undermining them, because with limited budgets and limited absolute control over many services, Metro's elected body depends more on brokering between interests than dominating them.

The electoral feature of Metro probably is its single most distinguishing characteristic. Although some may suggest that its electoral status allows Metro councilors to exercise greater powers than appointed boards, in fact Metro's powers are considerably less than those of many appointed boards. Metro's principal powers are in coordinating planning functions, providing professional analytic and planning services to local governments and special districts, and advising the development community on development opportunities within the region. An argument can be made that Metro is not the second tier of a two-tiered regional governance structure; it functions instead as an umbrella under which regional issues are addressed and consensus is built. Local governments and regional special districts retain virtually all powers of detailed planning, administration, and revenue production.

Although Metro's provision of services is narrow, it is a regional symbol of cooperation among otherwise competitive jurisdictions. The City of Portland long dominated operations of previous regional bodies, and its suburbs resented this. One of the concerns about the initial formation of Metro was that the executive officer would be Portland's high-profile former mayor and later U.S. Secretary of Transportation, Neil Goldschmidt. Metro's actual range of powers is so limited, however, that it is hard to imagine that anyone would view election as its executive officer as tantamount to being regional mayor. Perhaps this is why Metro's executive officers tend to be professional technocrats and managers, not really politicians. More to the point, the electoral structure of Metro means that Portland's regional influence has been diminished in favor of the populous suburbs—not that the suburban representatives themselves are unified in their view of how the region should be managed. Metro's governing body now represents a formalized structure for sharing power. This is perhaps the most important lesson of Metro. It is through a regionally elected body that regional interests are addressed and policies reflecting the regional interest are legitimized. In the meantime, the powers and resources of cities, counties, and special districts remain unaffected by Metro's presence.

The proposition is offered that as the lessons of Metro's structure and operations are learned, more "Metros" will be formed across the United states. What is that structure? In summary, it is a governing body that is elected by district with an executive officer elected at-large; that has a very low profile in the direct delivery of services, which in Metro's case is management of the regional zoo and landfills; that coordinates land use, facility, and transportation planning, including the allocation of federal transportation funds, in its role as an MPO; and that provides an umbrella under which regional issues may be addressed effectively.

NOTES

1. The Portland MSA is considered separate from the Portland (OR)-Vancouver (WA) Consolidated MSA. Vancouver is located north of the Columbia River, which separates Oregon and Washington in the CMSA.

2. For a discussion of edge cities, see Garreau (1991).

3. Public goods have qualities of nonexclusivity and nonexhaustibility. Parks—such as Washington Park, where the regional zoo is located—for example, find it difficult to exclude anyone from their use. If Portland pays for the park but suburbanites use it, suburbanites enjoy a "free ride." Air is nonexhaustible, but it is pollutable, to the detriment of everyone else breathing it. Services such as parks, zoos, and resource planning are considered public goods for which everyone must contribute; this is the underlying economic rationale for Metro.

4. Readers should appreciate that Portland is the nation's largest inland port, being roughly 120 miles from the Pacific Ocean at the confluence of the Columbia and Willamette Rivers.

5. One could argue that the Twin Cities have a three-tiered governance structure, but only if the Metropolitan Council's functions themselves compose two of those tiers.

6. Unlike the highly politicized processes seen in Denver, San Francisco, and Seattle.

REFERENCES

Abbott, C., & Abbott, M. P. (1991). *Historical development of the Metropolitan Service District.* Portland: Metro Home Rule Charter Committee.

Bureau of the Census. (1963a). *Census of Manufactures.* Washington, DC: Government Printing Office.

Bureau of the Census. (1963b). *County Business Patterns.* Washington, DC: Government Printing Office.

Bureau of the Census. (1987a). *Census of Manufactures.* Washington, DC: Government Printing Office.

Bureau of the Census. (1987b). *County Business Patterns.* Washington, DC: Government Printing Office.

Bureau of the Census. (1991). *State and Metropolitan Area Data Book 1991.* Washington, DC: Government Printing Office.

City Club of Portland. (1986, March 13). *City Club of Portland Bulletin, 66*(42). [Entire issue]

Garreau, J. (1991). *Edge cities.* New York: New York Times.

Leonard, H. J. (1991). *Managing Oregon's Growth.* Washington, DC: The Conservation Foundation.

Nelson, A. C. (1987). The effect of a regional sewer service on land values, growth patterns, and regional fiscal structure within a metropolitan area. *Urban Resources, 4*(2), 15-18.

Nelson, A. C., & Knaap, G. J. (1987). A theoretical and empirical argument for centralized regional sewerage planning. *Journal of the American Planning Association, 53*(4), 479-486.

Part IV

Conclusion

12

Perspectives for the Present and Lessons for the Future

H. V. SAVITCH
RONALD K. VOGEL

In this chapter, we attempt to provide some answers to questions posed in the first chapter, namely, what are the centrifugal and centripetal forces that shape our urban regions and their governance? Our answers are based on comparative analysis of the 10 city-regions chosen for their distinctive approaches to regional governance (see Chapter 1). These cases include four examples of metropolitan government—Jacksonville-Duval County, Minneapolis-St. Paul, Portland, and Miami-Dade; three cases of mutual adjustment—Washington, D.C. (interlocal government agreement), Louisville-Jefferson County (interlocal government agreement and public-private partnership), and Pittsburgh (public-private partnership); and three cases of avoidance and conflict—Los Angeles, New York, and St. Louis. First, we consider how each region responded to the challenges of regionalism. This discussion is followed by a set of lessons that can serve as a basis for understanding, treating, and developing policies for metropolitan regions by policy makers, local officials, practitioners, and academics.

■ Perspectives

We assess metropolitan governance in each community with attention to area coverage, functional scope and autonomy, planning and policy, and democratic representativeness (see Barlow, 1991). Area coverage refers to whether the regional coordinating body or mechanisms correspond to the functional city-region. Effective metropolitan or regional governance requires the existence of political institutions that are coterminous with the boundaries of the city-region. Functional scope and autonomy concern whether there is a broad set of functions over which the regional body has authority and whether sufficient authority, fiscal resources, autonomy, and implementation power are available to address regional problems. Policy and planning focus on whether the region can plan the community's growth and development, including infrastructure, land use, and economic development. Democratic representativeness refers to whether the regional decision-making institutions are based on direct election and represent racial, ethnic, income, and geographic interests in the region. Representativeness also relates to fairness or equity in the outputs and outcomes associated with regional governance.

Overall effectiveness of regional governance requires some attention as to whether regional bodies or processes exist to address or resolve regional issues in the first place. We will discuss separately each of the three sets of cases—avoidance and conflict, mutual adjustment, and metropolitan government.

■ Avoidance and Conflict

Los Angeles, New York, and St. Louis suggest that avoidance of regional issues and frequent conflict over economic development are associated with a high level of governmental fragmentation. Although this association is not invariable (see the discussion on mutual adjustment), it does occur, with damaging consequences for the larger body politic. Given the increased competition across the country, this pattern unfortunately may become more widespread.

Los Angeles

Saltzstein reports that for a short time, Los Angeles showed some movement toward regionalism, driven by the need to comply with federal air pollution standards. The air quality management board (AQMD) has had some success in getting localities to comply with measures designed to reduce air pollution, and there is a Southern California Association of Governments (SCAG). It is directed by elected officials from the counties, Los Angeles, and municipalities. This body is strictly voluntary, but it has gained some power over transportation policy, because it must approve local projects under the Intermodal Surface Transportation Act of 1991. Both of these regional bodies cover the five-county area of the region satisfying the areawide criteria. Thus, federal policies are responsible for the movement toward regional governance in the Los Angeles area.

On the surface, it appears that the range of issues and the authority of the AQMD are small, in that the services over which the two regional bodies operate are very narrow. The AQMD was able to gain some control over growth management and land use planning and policy making, and to set development policies in order to reduce air pollution. There was some early optimism of movement toward regional governance, if not regional government, as an outgrowth of efforts to address air pollution. The AQMD, however, fails as a regional government. It has not been able to enforce its plans, and local governments successfully have resisted its regulations by co-opting board representation.

A regulatory model of regionalism undertaken by technicians and bureaucrats insulated from democratic control has had some success. In the long run, however, this may be a hollow victory, because no other regional strategy or institution is offered in its place. Given economic competition among localities, a continuing recession, and suburban efforts to separate themselves from the inner city, it is difficult to see how regional problems can be addressed in the long term.

New York City

Kantor and Berg's study of New York City paints a picture of avoidance of regional governance and conflict among subgovernments. The region comprises 24 contiguous counties in three states (New York, New Jersey, and Connecticut) that together make up the Consolidated Metropolitan Statistical Area (CMSA). With a population of more than 18 million, the CMSA does not have an all-purpose body to coordinate strategic planning or manage economic development.

Competition for industry often is intense. The states of New Jersey and Connecticut and localities within them frequently are accused of raiding New York City to attract business. Newspaper advertisements tout the advantages businesses could gain by moving across state boundaries, economic development specialists openly solicit corporations, and government officials outbid one another in trying to attract sports franchises.

Despite difficulties, the region has forged some limited ventures for cooperation. Most notably, the Port Authority of New York and New

Jersey successfully manages the airports, bridges, and some subway lines, which are the region's commercial arteries. The Port Authority also has invested in profitable real estate (the World Trade Center). Add to this the mass transit system (Metropolitan Transportation Authority or MTA) and economic development (New York State's Urban Development Corporation or UDC), and we can discern a quilt of cooperation in the form of ad hoc agreements.

The overall pattern, however, is clear: The New York CMSA does not have regional governance and shows little sign that greater cooperation is possible in the foreseeable future. The Port Authority, MTA, and UDC perform planning and development functions within narrow realms, but no regional body exists to plan or coordinate. There also is evidence that postindustrial change will result in continued population and business dispersal.

St. Louis

Phares and Louishomme point to St. Louis as the ideal example of avoidance and conflict. There is no sense of a functional city-region. Even a limited conceptualization of the region as defined by the county boundaries is absent. Having voted to separate themselves in 1876, both city and county bear the name of St. Louis, but no further affiliation exists. St. Louis is one of the most fragmented regions in the country. The fragmentation is matched by considerable discrepancy in the property tax base. This disparity confirms a pattern of mounting class and ethnic imbalance that inhibits the region from building cooperative alliances.

There have been attempts to establish regional governance, but these have been either thwarted by the courts or turned down in referenda. The region has been caught in an internecine contest for economic development, especially between the city and county.

These tensions have been especially acute in struggles over tax receipts. The importance of this revenue stream led the region's numerous municipalities to preempt one another in imposing sales taxes. To quell the fiscal wars, the county adopted a measure under which localities could choose between "point of sale" and "pool" taxes (tax sharing). A tax war followed. The debate over sales taxes is replicated by similar quarrels over the distribution of utility and gaming taxes.

Amid this competition, some cooperation does take place, mostly in transportation and economic development. An interstate compact between

Missouri and Illinois led to the creation of a Bi-State Development Agency covering seven counties. An East-West Gateway Coordinating Council is responsible for promoting transportation, and there are regional programs for stimulating economic recovery.

These are relatively modest efforts, but they do point up the power of economic necessity. Overall, the St. Louis experience bears a resemblance to that of New York. Both demonstrate the centrifugal dynamics that stem from existing fragmentation, which is coupled to economic and social disparity. Both regions also show some resilience in adopting limited measures for cooperation, but nevertheless both exhibit a scanty record of regional effectiveness.

■ Mutual Adjustment

Interlocal agreements and public-private partnerships can be treated as cases of mutual adjustment, wherein public and private actors find ways to address issues of regional concern without the creation of formal metropolitan government. This approach is consistent with the public choice position that local governance is best understood as a complex system. Political fragmentation encourages competition among cities that leads to efficient and effective service production. In services for which economies can be realized if provided on a larger scale or on issues of metropolitan or regional concern, coordination can occur without resorting to centralized governmental structures. Three of our cases illustrate the mutual adjustment approach to metropolitan governance.

Louisville

After two failed merger efforts, the City of Louisville and Jefferson County adopted what has come to be known as the Compact. This interlocal agreement represents a comprehensive effort that includes tax sharing of the local income tax (occupational tax), a re-sorting of services between the city and county governments, and a freeze on annexation for the duration of the compact. Under the terms of the compact, a number of agencies remain jointly operated, and a new joint agency, the Office of Economic Development, was created. Joint services that are provided countywide include economic development, transportation, libraries, and sewers.

Interlocal agreements of this sort also accommodate public-private partnerships. The mayor and county judge (chief executive) also have entered into a formal partnership with the business community in economic development. The city and county helped finance the creation of the Partnership for Greater Louisville, which provides incentives for business, underwrites economic development, and recently has supported regional strategic planning.

Jefferson County covers nearly 70% of the metropolitan area. The existence of the Compact and strong public-private partnership in Louisville and Jefferson County provide a reasonable degree of administrative integration that enhances regional cooperation. The Compact covers a broad range of services that are provided jointly and also provides for jointly appointed advisory boards for other services that are carried out by one or the other government. The compact was not subject to a referendum but was approved by the City Board of Aldermen and county Fiscal Court as well as the state legislature. The public-private partnership and a new comprehensive planning process are incremental ways of addressing regional economic needs.

The Compact has worked fairly well, although there are tensions between the needs of the inner core (held by the city) and concerns of the suburbs (held by the county). The county is unhappy with the tax sharing agreement that has led the city to receive an additional $20 million between 1986 (when adopted) and 1995. The compact is set to expire in 1998, and although agreement is probable, failure to achieve an accord could reignite city-suburban competition.

The major limitation of the Louisville approach has been that it does not cover the bi-state region (Southern Indiana), which includes six other suburban counties with about 1 million persons. The only regional institution there is the Kentuckiana Planning and Development Agency (KIPDA), which acts as the metropolitan planning organization, areawide agency for the aged, and a state planning agency. There is no council of government or other forum for regional cooperation. The Compact and Partnership are focused in Louisville and Jefferson County. It is not surprising that the different interests generate interstate tensions. Thus, when the Partnership attempted to undertake a regional economic development strategy, Southern Indiana resisted. Its own demands for a new bridge across the Ohio River (a distance away from the City of Louisville) were not supported. This is a serious impediment to regional cooperation.

Washington, D.C.

A council of government approach to regional cooperation and governance is pursued in the Washington, D.C., metropolitan area. The Metropolitan Washington Council of Governments (MWCOG) is a multipurpose regional body that provides a forum for interlocal cooperation and coordination. The council has a large staff and budget, financed by membership fees, contracts, and grants. Its area coverage includes the District of Columbia and 15 cities and counties in Virginia and Maryland. Although the MWCOG covers only part of a larger PMSA and CMSA, its coverage is fairly comprehensive.

The MWCOG's authority and power are quite limited. It is entirely voluntary, and local governments have withdrawn or threatened to withdraw. It has no taxing powers or regulatory power. Its primary power results from its position as the metropolitan planning organization for the area. Its real strength comes from its planning capacity, knowledge base, research expertise, and ability to provide selective services to members. The council also serves as a clearinghouse and recipient of grants that local governments would not otherwise obtain, and it has been a resource for cooperative purchasing.

The council has no power to engage in broad, comprehensive planning, and it has been limited in its ability to carry out strategic planning in the region, although it is active in planning for transportation and law enforcement. Its primary limitation seems to be that it cannot afford to offend member governments. This prevents it from serving as an effective forum.

The MWCOG illustrates the public choice position that arrangements for providing and producing services and for coordination in metropolitan areas can arise without formal metropolitan government. The metropolitan council, however, is not a replacement for regional government, and it is unable even to serve as a forum for regional issues.

Pittsburgh

Pittsburgh is an example of the public-private partnership approach to regional governance. The partnership between public officials and the private sector is focused in the city of Pittsburgh but also takes in Allegheny County government. Areawide coverage is fairly good.

The partnership is not located in a single organization but has sustained itself over several decades. It has managed the transition of Pittsburgh from an industrial city to a postindustrial city by focusing on redevelopment, a diversified economy, and a reliance on high technology. The partnership has covered a broad range of issues (functional scope), creating commissions and nonprofit organizations to carry out its programs and acquire funds from public and private sources to pay for its projects. The partnership takes a strategic approach to issues and plays a strong role in planning the economy and physical redevelopment. Its agenda for social development, however, is stunted.

The main weaknesses of the public-private partnership strategy are the lack of democracy and an almost exclusive attention on business development. Successful restructuring in Pittsburgh involved thousands of people leaving the area when they could not find work as manufacturing declined. Few social programs or human development strategies were pursued to address the plight of the unemployed, and redevelopment policies were not always sensitive to neighborhoods.

■ Metropolitan Government

There is a great deal of variation in what constitutes formal metropolitan government. Our cases demonstrate both the promise and the disappointment that metropolitan governments pose.

Miami-Dade

Metro Miami-Dade, although embodying the ideals of the two-tier metropolitan government model, does not adequately cover the city-region. Metro-Dade governs only a county within the larger region of South Florida, which encompasses at least Broward, Dade, and Monroe Counties. Even the South Florida Regional Planning Council does not include Palm Beach County. It is interesting that the two-tier model of governance has been compromised within Dade County, as a majority of the population now lives in the unincorporated area, with the Metro government providing local and areawide services.

Metro-Dade does provide a broad range of areawide services within the county boundaries, including solid waste disposal, sewerage, parks and recreation, land use planning, growth management, and police and fire protection. Furthermore, the metro government has the ability to set minimum standards of services by cities; it can adjust city boundaries and incorporate or disestablish cities. The only major power that Metro lacks is in zoning.[1] Metro-Dade does have the capacity to engage in planning for its share of the region through its growth management process, as mandated by the state of Florida. It also is directly responsible for the administration of most of its programs and has the power to set service standards, even in the municipalities, to further the implementation of its policies.

Metro-Dade was governed by a seven-person commission, including an elected county mayor, and operated under the county manager form of government. Thus, there was direct election of the political leaders. However, concerns of representativeness were raised by minority-group members critical of the at-large election system. The county recently settled a voting rights suit enlarging the commission to 13 seats and shifting from at-large elections to single-member districts.

Effectiveness of the Metro-Dade system is difficult to gauge. On one hand, it has provided professional government in the county since its inception in the 1950s and led to the provision of a modern infrastructure and orderly development of the community. In more recent years, however, the system has been criticized for failing to maintain a "two-tier" model of service provision, for a lack of leadership, for failing to address social problems, and for shortcomings in responding to issues of ethnic and racial diversity that were manifest in repeated riots.

The major reform currently under consideration is the creation of a strong mayor to ensure greater focus and executive leadership. Metro government is handicapped in that its authority is limited to the county boundaries (i.e., a "regional" county rather than a metropolitan government) and by the fact that there are several other regional agencies (the South Florida Water Management District and South Florida Regional Planning Council) that carry out planning within the broader region. Furthermore, Metro does not have the means to address broader regional problems, and no other body exists that integrates the full South Florida region politically. Finally, the bureaucracy's professional ethos and management style may no longer be compatible with a changed political culture that accompanied the dramatic growth in the Hispanic population and increasingly volatile racial situation in the metropolitan area.

Minneapolis-St. Paul

The Twin Cities Metropolitan Council was designed specifically to be a regional government, covering a seven-county area encompassing the twin cities of Minneapolis and St. Paul. This distinguishes it from county-oriented systems (such as Miami-Dade and Jacksonville), which cover only a portion of their true metropolitan regions. Until recently, Minneapolis-St. Paul followed a bifurcated model. Its Metropolitan Council was responsible for regional planning and policy concerning transit, housing, land use, and other issues of metropolitan significance. Implementation was carried out by metropolitan commissions, other agencies, or local governments.

The metropolitan council is a general-purpose government with considerable authority to plan the region's infrastructure. This seemingly broad grant of authority is undermined, however, by the lack of direct elections—the council is appointed by the governor. This lack of an independent power base, coupled with an erosion of political support by other constituencies, restrains the council from taking a leadership role in many important development issues over which it has authority, including the construction of a domed stadium and the largest mall in the world. The bifurcated model also provided a means for resistance to the metropolitan council's policies, as it had to work through other groups. This has led to inefficiencies and stalemates in certain areas, such as transportation.

The metropolitan council's greatest accomplishments were in its first decade, when it developed a regional system for sewage treatment, developed regional parks and turned them over to the counties, and authored a metropolitan development plan to guide urban growth and infrastructure. The state legislature incrementally added to the metropolitan council's stature and powers through the early years. The tax sharing feature of the metropolitan government is still the envy of many communities.

In the last decade, the metropolitan council has fallen on hard times. It frequently is bypassed on major development initiatives, which undermines its planning. It has had little success in social planning, reducing income disparities between central cities and the suburbs, or preventing exclusionary zoning in affluent suburbs. The major obstacles to effective regional government today appear to be lack of direct, popular election, weak leadership, an insufficient political base, and a separation of policy from implementation.

In the last year, the state legislature has considered and adopted some new reforms to strengthen the metropolitan council. It discarded the bifurcated model, providing for the metropolitan council to administer directly programs such as transportation. This led the legislature to create the role of chief administrator, responsible for administration. This should strengthen the metropolitan government's leadership role. The council also was made more dependent on the governor, who now has the power to remove as well as appoint members. Moreover, the legislature has not shown an inclination to press for further equalization and refused to reduce barriers for low- and moderate-income housing in the suburbs.

Jacksonville-Duval

Jacksonville-Duval, a consolidated city-county, governs only one of four counties in the metropolitan statistical area. Thus, Jacksonville illustrates the problem with city-county consolidation as a strategy to achieve metropolitan governance. Still, Jacksonville-Duval contains just under 75% of the metropolitan area population. Using a city-region criterion, the city-county consolidation is not a sufficient areawide boundary. Regional planning cannot take place effectively because much of the growth and development that affects the region is outside the control of the consolidated city-county government (Self, 1982). Furthermore, the consolidated city-county cannot annex further into the hinterlands, either to plan regional growth or to incorporate urbanized areas into a regional governance structure.

Jacksonville-Duval, as a combined city-county, has only modest authority in urban service delivery. Although it acts as a multipurpose general government that provides essentially one level of service area-wide, several affluent suburban cities were excluded from the consolidation[2] and by the continued existence of a number of independent authorities including transportation, downtown development, and a port. These public authorities account for about 40% of local spending.

The consolidated city-council is governed by an elected mayor and 21-person council operating under a strong mayoral form of government. Race was a major factor in why the two governments were consolidated; the African American community would have been a majority of the old former central city. Although the council is directly elected, consolidation did result in minority dilution, and participation rates in elections seem to have fallen.

With respect to effectiveness, there is no doubt that the consolidated city-county has provided a modern and efficient infrastructure for development in the former suburban parts of the community. The former central city has been less successfully redeveloped. The consolidated government is more professional and has been more willing to use federal and state programs to pursue redistributive policies.

The suburban interests clearly have dominated and benefited by the merger of the two governments. The inner city continued to decline, and segregation and income disparities have increased. On balance, the city residents appear to be subsidizing fire protection, parks and recreation, water and wastewater treatment, libraries, and development infrastructure in the suburbs. In return, the city residents gained more equal property assessments and continue to receive free garbage collection. The public continues to support the consolidation, but citizen satisfaction with leadership has declined, as reflected in the adoption of term limits and a cap on property assessments.

Portland

It would appear that the Portland Metro has substantial areawide coverage, with authority over three counties that make up the urbanized area of the metropolis. (It does not cover the Portland-Vancouver CMSA or the full Portland MSA.) The Portland metropolitan government is not, however, a full general-purpose local government. Still, this government may eventually grow into a fuller multipurpose government. It is governed by a seven-member council elected by single-member districts in nonpartisan races for 4-year terms. An executive officer is elected at large and is responsible for administration. This direct election provides a strong political base for leadership. Metro has the authority to assume management of regional bus and light rail system. The metropolitan government also has a solid fiscal base, with the ability to levy sales and income taxes, subject to regional referendum.

To date, the primary role played by Portland's metropolitan government has been the planning of the region's development, including growth management, land use planning, and transportation planning. Nelson indicates that it is best understood as an umbrella agency whose primary role in the region is the coordination of planning. It is certainly not a full-service regional government, but it is well suited for planning, in that it serves as the metropolitan planning organization for the region. It also provides a forum to address regional issues. There is some evidence that Metro has reduced the City of Portland's regional influence in favor of suburbs, but it has also provided a "formalized structure for sharing power." Because its role is largely planning, it is not viewed as a competitor government by other local governments, facilitating its role as mediator among regional interests. On the other hand, although it has an independent political base, it has little direct power to provide services or implement its policies. This means that it can address only those regional issues on which there is a substantial consensus.

■ Lessons

A major aim of this volume is to build an empirical base on which to revisit questions of metropolitan and regional governance, a field in which normative predispositions and myths have obfuscated more than elucidated the choices facing urban regions. The cases chronicle the efforts of public and private officials to meet the challenges of regionalism. Leaders struggle to make their regions competitive in the global marketplace, to secure a sound economic and financial status, and to attract jobs and provide a better quality of life. At the same time, they seek to address severe inner-city problems and to meet the demand for infrastructure placement and services in the suburbs. Regional strategies are shaped in most cases without benefit of general-purpose regional governments.

So what chance do regions stand without governments? We began this volume with the promise to examine the experience of metropolitan regions, evaluate how they have evolved, and identify forces of resistance and cooperation. We now put this experience in perspective by offering observations about regional composition, dynamics, and strategies. Each of these factors is meant to cover the political economy and scope of our regions (composition); the process of interaction and the opportunities, risks, and changes that regions confront (dynamics); and the manner by which regions exploit possibilities for success (strategies). We couch these observations as seven basic lessons that can serve as a basis for understanding, treating, and developing policies for metropolitan regions.

Lesson 1: Economic Complexities Yield Different Modes of Political Response

Our survey of 10 regions shows that regions have become economically more differentiated and more complicated, but also more closely coordinated. As industry has decentralized, regions are linked inextricably through transportation, communication, and the functional divisions. It is not uncommon for corporations to establish downtown headquarters, keep back offices in suburbs, and maintain warehouses still farther away. The same is true for universities and hospitals, which retain established centers in the central city and branch campuses in the suburbs. All of this necessitates increased interaction by those who work and reside in different parts of a region. Whether this is done by routine phone calls, periodic visits, or instant communication via the information highways is immaterial. The fact is that economics and technology have joined to produce "functional cities" that stretch throughout metropolitan regions.

This is illustrated in how and where people earn their living. The introductory chapter shows increased commutation between central cities and suburbs, but it is important to remember that our major story is one of intraregional interaction. People, on average, earn their living farther away from a primary residence than ever before. More money flows out of central cities into suburban areas—and more money flows between suburban areas.

Figure 12.1 shows this pattern for each of our 10 regions. The bar graphs highlight the monetary flows by people who live in a suburban county but work in the central county (in-commuters) as well as those who live in one suburban county but work in another (out-commuters).[3] Negative flows, below the base line, indicate money flowing out of an area, whereas positive flows, above the baseline, indicate money coming into an area.

The economic interdependence of cities and suburbs is evident over the 20-year period between 1970 and 1990. Every central county was below the baseline and had commuters in net taking money out of those areas and into surrounding suburbs. Just as important, 9 out of 10 suburban counties were above the baseline and had a net flow of money coming into them from outside.[4] Between 1970 and 1990, the bar graphs increase markedly upward or downward. Jobs and payroll increased within metropolitan regions—whether seen through increased flows out of central counties or increased financial linkages between suburban counties. It is fair to surmise that this trend was complemented by other transactions in mortgages, investments, and commercial arrangements.

Were these forces irrepressible? Certainly, centripetal pressures cannot be ignored—transportation needs have to be accommodated, labor has to be trained, and capital investments must be recruited. Regions handle these pressures in different ways. Regional responses to these pressures bear a certain relationship to the level of political integration within a particular area. Political integration may result from centralization in the public sector (city-county consolidation), federated regions (two-tier metropolitan government), the presence of elite networks or power structures (public-private partnership), or a combination of forces (Vogel, 1992).

More unified areas—Jacksonville, Minneapolis-St. Paul, Pittsburgh, and Louisville—show the widest and most explicit embraces of the

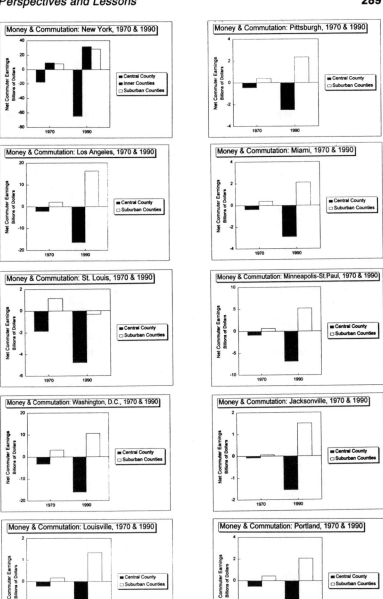

Figure 12.1. Flows of Commuter Earnings, 1970 and 1990

challenges of economic development. Swanson reports that Jacksonville's surge of economic development came about because formal consolidation of city and county governments enhanced the ability of the "growth machine" of developers, bankers, insurance companies, and lawyers to implement their growth agenda, leading to everything from professional sports to a headquarters for American Express. Minneapolis-St. Paul used its MetCouncil to channel growth and control development, at least during its formative years. Planning and the containment of urban sprawl were hallmarks of that region during the 1970s. Harrigan concludes that under current pressures, MetCouncil may be shifting its energies toward stimulating economic growth and keeping up with the demand to produce jobs.

In Pittsburgh, Renaissance I, II, and III forged public-private partnerships to expand, convert, and diversify the regional economy (from "steel city" to "software city"). The mechanisms are ad hoc, though powerful. They include interlocking regional networks to finance development and business linkages to research facilities—all sustained by a host of surveys, technical assistance, and strategic plans. This occurred even though Pittsburgh is renowned for its extreme fragmentation of local government.

Louisville relied on a comprehensive interlocal agreement between city and county governments (the Compact) that unified the office of economic development, placed a moratorium on annexation, and allowed for tax sharing. This was followed by the creation of a formal public-private partnership that forged consensus on a regional economic development strategy including an expanded airport.

Fragmented regions are more selective and circuitous. New York presents a classic use of narrowly gauged functional agencies to build infrastructure and promote development. These agencies (a bi-state port authority and a state development corporation) are self-sustaining, insulated from democratic control, and able to act quickly and flexibly. Kantor and Berg show that in spite of regional animosities, bi-state agencies still manage to forge regional links and are still resilient. St. Louis also follows in this mold. Phares and Louishomme illustrate how metropolitan and bi-state agencies seek to enhance development and undertake planning and transportation functions.

Still another category is that of mixed, technical cooperation. Here, the centripetal pressures are not directly economic. A few regions seem content to focus their energies on the coordination of information. Washington's MWCOG concentrates on activities that all constituents can get behind, such as crime control. Portland's Metro took up issues that localities were only too happy to offload (the zoo, sewerage, and transportation planning). Los Angeles has sought regional cooperation on an issue (air pollution) that drew immediate fire but whose long-term resolution is technical. In Miami-Dade, water management and planning, under a framework established by the state, set the parameters for cooperation in the larger South Florida region.

Lesson 2: Look Behind the Formal Trappings of Government

Not too long ago, a colleague suggested that we might resolve the debate over which is better—metropolitan or fragmented forms of government—by putting their respective performance to a hard empirical test. Why not, he advised, go directly to the heart of the matter? Select areas that have adopted "metropolitan government," choose performance indicators, and systematically evaluate results as compared with more fragmented counterparts. The solution seemed easy, simple, and direct—except for the thorny facts.

The truth is that metropolitan government is not a singular form. As demonstrated in these cases, "metros" can include multiple or single tiers. Even within supposedly common forms, the variations in design, scope, function, and power are enormous. Miami, Minneapolis-St. Paul, and Portland are formally "two-tier governments," yet the resemblance does not go beyond formalities. Miami's "metro" is confined within a single county, whereas Minneapolis-St. Paul's "metro" extends across seven counties and Portland's covers most of the MSA. Metro-Dade is designed to deliver areawide services, Minneapolis-St. Paul is geared to equalize regional burdens, and Portland functions as an "umbrella" responsible for specialized functions.

Jacksonville has an altogether different form and content. Its "metro" lies within a consolidated county. In this respect, Jacksonville is more akin to Metro-Dade—though it is nontiered and elects a common council, which also puts it closer to Portland. Stowers points out that the majority of the population in Dade County lives in the unincorporated area and is served by one tier, the county. Furthermore, Swanson points out that consolidation in Jacksonville also led to the creation of several new independent public authorities and left several suburban cities untouched. Is this more centralization or less? Once examined, the picture is complex. Metropolitan government may be the common genus, but it has many species. We recognize the distinct characteristics of each "metro" government and appreciate that one or two labels do not suffice to cover the variation.

In looking at regions characterized by "mutual adjustment," we learn that form does not always equal function. By all accounts, Pittsburgh is one of the most fragmented regions in the country. Its home county has more than 300 governments serving just 1.4 million people and includes 12 cities with fewer than 1,000 residents each. Under the fragmented form, however, lies a unifying net of business elites that ties the region together through public-private partnerships. Jezierski's chapter illustrates how this elite worked for more than half a century to restructure the economy, attract investment, and rebuild the city.

Louisville also is nominally quite fragmented, with more than 120 governments within its home county. There, formal fragmentation was overcome by interlocal agreement and public-private partnerships. Economic integration followed political integration. Researchers using the Census of Government would not recognize the degree of political integration that has occurred in the county.

As a region characterized by avoidance and conflict, New York adds stills another anomaly. A century ago, it became one of America's first "regional governments" when it merged five counties under a single municipal government. Today, it is counted as a central city, fighting "place wars" with surrounding suburbs and caught in the throes of local conflict. New York manages to extend beyond its borders and cooperate along selective, ad hoc lines. Much the same story can be told for counterpart regions—St. Louis and Los Angeles. Selective, ad hoc pacts carry out necessary functions and allow for limited cooperation across boundaries. The fact is that localities need their neighbors.

Lesson 3: Regions Work Out Cooperative Patterns in Particular, Least Controversial Ways

We can best understand regionalism by recognizing that it is made out of politics. The regional terrain is traversed by navigating its bridges and fissures. Coalitions and splits between groups make certain kinds of regionalism possible and preclude others. The process of eliciting cooperation creeps along slowly; it is incremental, and it is based on trial and error. Generally, solutions are negotiated around obstacles, so that thorny problems are avoided.

"Easier" or more complete instances of regional cooperation are found in Portland and Miami. In each of these cases, regionalism is built on consensus issues. For all of its homogeneity and popular support, Portland's metro is a conservative and modest enterprise. As Nelson demonstrates, Portland's metro has no power to control the localities, and its primary authority is geared toward planning, transportation, landfills, and zoo management.

The much heralded Miami experiment has a marginally broader scope, though it also has acted cautiously on cutting issues. Miami-Dade started off with power over local zoning but in a hail of controversy gave it back to the localities. Although it can still influence local land use through regional planning, metro is careful to rely on state authority to do so. Otherwise, Miami has stuck to tending to parks, sewage, and technical services. On paper, Metro-Dade has a great deal of local power and autonomy; in practice, it has been reluctant to exercise these prerogatives because of the firestorm that would follow.

Regions that have attained moderate cooperation shy away from controversy and instead build on common agreement (Washington, D.C.). When regions do confront cutting issues, they are dealt with piece by piece or managed by public-private partnerships (Louisville and Pittsburgh). Washington's MWCOG is formulated around everyone's common foe— crime. Henig, Brunori, and Ebert demonstrate how the MWCOG is relegated to information sharing and draws funding from outside sources. Regionalism in Pittsburgh is largely contingent on Allegheny County covering for local jurisdictions, private and semipublic institutions that pick up the slack, and elite networks that promote development.

Even the most strained cases of regional cooperation—those ridden with avoidance and conflict—have tacked around the obstacles and fashioned their own institutions. New York's Port Authority is a massive and powerful institution. It has managed to survive on an internally generated budget and with insulation from popular control. St. Louis and Los Angeles have done likewise—either turning to nonelected bi-state agencies (development in St. Louis) or by relying on special districts and county supervisors to carry out necessary functions (improving air quality in Los Angeles).

Obversely, regional cooperation has been set back when it directly confronts cutting issues. Minneapolis-St. Paul's MetCouncil went "sour" because of lingering doubts about housing equalization and the redistribution of financial resources. A conservative governor, untrusting legislators, and right-wing interest groups put regionalism's most bold experiment into question. Turning back to St. Louis, efforts to establish metropolitanwide authorities and redo local boundaries were defeated amid a debate over cutting issues. Fiscal disparities, local taxes, zoning, and racial segregation were the hot buttons of regional opposition, and they eventually frustrated government reform.

Does this mean that regionalism—or partial variants of it—must count on innocuous issues to succeed? Probably so and usually true, but the answer extends beyond a few simple words. Regionalism is not an end result but an incremental and evolutionary process. It builds on successes, relapses upon failure, and shifts strategies with the times. Given a large amount of popular backing, regional institutions can succeed in handling hot topics. Minneapolis-St. Paul's MetCouncil rode high during the 1970s, and its policies on land use and social equalization were acclaimed. Twenty years later, MetCouncil was regarded differently by its overseers and its constituents. It could no longer make the same bold decisions, and because of this, it became a different institution.

Louisville's Compact addresses conflictual issues—tax sharing, annexation, and functional consolidations—but these were resolved only after the more sweeping reform of city-county consolidation was taken off the table when voters rejected it twice. In addition, the prospect of intensified place wars was more troubling than forging cooperative relations in a community suffering the effects of deindustrialization. The stage for larger cooperation and compromise was reached by incremental steps toward cooperation, achieved over several decades, including joint agency operation and the external threat of state legislative intervention.

Lesson 4: Regionalism Addresses Different Problems in Different Regions

Our cases show different problems arising in different areas as well as differences in priorities. One-size solutions do not fit all regions. Thus, in Louisville, the problem initially was a declining city fiscal and economic base. The solution turned out to be not merger but tax sharing and functional consolidation. Regional cooperation did, in fact, address downtown decline, and it helped in its partial rehabilitation. In Pittsburgh, the problem was deindustrialization, and the solution was a public-private partnership to manage the transition to a postindustrial economy. In Washington, D.C., the problem was to provide a forum for regional decision making; the Council of Governments filled the bill. In Miami, the problem was rapid growth and how it overwhelmed a rural, antiquated county government's capacities. In Jacksonville, the problem was failed services associated with a corrupt, entrenched political machine. Consolidation became a reform mechanism and successfully enhanced local services. Portland Metro was created to provide regional planning and absorb certain services. Minneapolis-St. Paul was set up for fiscal equalization and to rationally plan the region. The AQMD was set up in Los Angeles to address a single problem, air pollution, with regional implications.

Lesson 5: Although Regionalism Can Be Managerially Viable, It Is Politically Fragile

A colleague once quipped that "a metropolitan region is that jurisdiction larger than the one for which we cannot find an answer." The jest touched on an all too familiar effort to kick responsibilities onto different levels of government in order to solve intractable problems. The irony springs from the fact that although regions theoretically may encompass an optimal scope for issue management, they lack the political clout to do anything beyond managing policies made elsewhere. Regions may be managerially competent, but, by and large, they are politically weak, and this makes all the difference when it comes to taking on bold policies.

The reasoning is simple. Most regions lack a loyal and dedicated constituent base. Smaller jurisdictions engender strong identities and stalwart defenses. The cities and suburbs that compose metropolitan regions unabashedly act in their own self-interest. They annex land (Louisville and St. Louis) try to impose taxes on commuters (Washington, D.C.), pirate business from their neighbors (New York), and defensively incorporate to keep outsiders away (Los Angeles). At the other end, states have supreme legal authority, enjoined by the Constitution to take all residual powers and act as sovereign entities. In between the localities and the states are the regions. They thus appear destined to be ground between two mightier partners.

Our cases bear out this proposition. Portland's metro survived because it brokered agreements between the localities and took on functions they never wanted. Nelson notes that Portland's metro is more an "umbrella" that brings interests together than a "government" that directs and commands. Washington's MWCOG is a quintessential case of a coordinating body that quickly backed away from taking controversial positions (gun control, commuter taxes, or statehood for the District of Columbia). On more than one occasion, it almost collapsed over threats that members would withdraw.

At their most extreme vulnerability, regional bodies impose costs, quickly acquire enemies, and have no constituency on which to rely. Saltzstein shows how this worked in Los Angeles, as hostilities broke out between the Air Quality Management District (AQMD) and local officials. The AQMD clearly was at a disadvantage and eventually had to relent. In another instance reported elsewhere, the AQMD is alleged to have sent a form letter to corporate executives, threatening to fine or imprison them for filing faulty car-pooling plans. The letter was not only a clumsy attempt to deal with a problem but also belied the agency's credibility. One critic complained, "What kind of regional government is that?" (Rabinovitz, 1994).

One might counter that regional institutions are weak only because they possess no strong legal foundation, but this begs the question. Regional governments lack legal foundation because they are politically squeezed out, lack a definitive constituency, and find themselves in a political netherland. Jacksonville appears to be an exception that proves the rule. As a metro government, it is coterminous with a county and draws strength from a single, unified jurisdiction. Jacksonville, then, can afford to act vigorously, flex its muscles, and become the prize of the local growth machine because it has no real competitors.

Even in Jacksonville, fragility is apparent. Unity was created by excluding some parts of the community, the suburban cities, and disenfranchising another part, the old central city, home to a large minority population. Furthermore, it is politically infeasible to extend the consolidated city-county regional boundaries to include the three remaining urbanized counties that make up the metropolitan area and that will pose great regional challenges in the future.

There are two other political problems facing regional institutions. African Americans believe that going regional will deprive them of black mayors and legislative power. White suburbanites believe that going regional could sap their control over land use and threaten their exclusivity. The only major and consistent interest group supporting regionalism is business—mostly because regionalism is well suited to manage economic development. It is no accident that metropolitan initiatives often are run by chambers of commerce or business associations (New York,

Louisville, St. Louis, Pittsburgh, and Jacksonville). This support, however, rouses its own reaction and populist ire against alleged business domination.

Regionalism's fragility can be attributed to its multiple adversaries and narrow political turf. Although managerially viable, its political logic is weak—mostly because it offers little direct, tangible, and immediate payoff to enough constituents. Grassroots interests distrust it, big government is wary of its intrusions, and it has too few beneficiaries. It is no wonder that Minneapolis-St. Paul has problems, that Miami metro covers an increasingly smaller part of the area, and that Portland is so cautious.

Lesson 6: Regionalism Operates at the Seams of Other Governments and Should Build on Its Strategic Position, but Cooperation Should Not Be Mistaken for Governance

Regionalism is a child of other governments. Federal largesse promoted councils of government (COGs) in the 1960s, and many of these survive today. More recently, clean air legislation and the Intermodal Surface Transportation Efficiency Act (ISTEA) have reinvigorated regional bodies, such as Metropolitan Planning Organizations (MPOs), to coordinate transportation planning (Federal Transit Administration, 1995). States, counties, and cities also spawn regional bodies. The range is considerable and includes bodies ranging from New York's and St. Louis's bi-state agencies all the way to Louisville's Compact and Pittsburgh's Allegheny Conference.

These actions put regional authorities, most lacking formal governmental authority, in an awkward position. They must act at the behest of others, work tactfully at the margins, and splice together pieces of authority. At times, they are given the dubious distinction of imposing onerous legislation or decisions on vocal interest groups. It is not surprising that regional bodies seek a more tenable role by preferring to offer benefits rather than exact costs.

Survival has its price. More and more, regions cited as viable are those that pursue strategies of "mutual adjustment" rather than formal "metropolitan government." Scholars and journalists alike applaud interlocal agreements and public-private partnerships because of their noncontroversial and flexible features. The voices of approval have now become a chorus, chanting that some regions have the best of all possible alternatives and enjoy "metropolitan governance without metropolitan government" (Ehrenhalt, 1989; Parks & Oakerson, 1989; Advisory Commission on Intergovernmental Relations, October, 1993; Peirce, Johnson, & Hall, 1993). There is little doubt that the phrases have appeal and that "mutual adjustment" does have advantages, but its proponents mistake bureaucratic cooperation for governance and, in doing so, obfuscate the meaning of government.

"Governance without government" is an oxymoron. Pittsburgh may have arrangements for routine services and limited regional cooperation; it does not, however, have regional governance. Disaggregating individual jurisdictions so that they can conclude interlocal agreements does not create regional governance, any more than common markets create nation-states. There is no getting away from the essentials of a common jurisdiction carrying out public policy, and regions do pay a price for not having that kind of institution. Part of the toll can be seen in the endemic "place wars" for business investment and capital, in threatened annexations and land rushes, in broken "nonaggression pacts" between states, and in deficient mass transit and environmental degradation.

Neither have strong metropolitan governments demonstrated that they can solve these problems (Barlow, 1991; Rothblatt & Sancton, 1993; Self, 1982). Our cases show that its partial variants have made some achievements, but not to a much greater extent than cases of mutual adjustment.

The situation is compounded by a tension between policy achievement and political viability. Herein lies a regional paradox. If metropolitan regions are to pursue effective policies, they must be politically viable (i.e., command popular and elite consensus), yet regional bodies whose policies go beyond the bounds of consensus are apt to lose that viability. In effect, the more aggressive regions become, the less power they possess. Regional bodies then must forever balance these tensions, trading off and adapting themselves to pressure and circumstances. The challenge is to do this while taking a long view of the need to convert political legitimacy into broader policy mandates.

Lesson 7: Regional Strategies Are Best Promoted By Unobtrusive Support, Positive Incentives, and a Long-Term Process

More than most governmental forms, regional bodies are apt to grow, change structure, and adapt to local change. Although less tenable, regions also are more flexible than other forms of local government. Federal, state, and local authorities should carefully and unobtrusively play to these strengths. Keeping in mind that each region has its own assets, liabilities, and pressure points, regional governance can best be built through its own momentum. This approach promotes a process rather than a definitive or singular end.

As a general strategy, it would be prudent to work from sources of support rather than test routes of federal or state imposition. Generally, legislatures are made up of the very people who are likely to resist, and the imposition of regional governance rarely wins their support. Instead, we suggest that regionalism start on a process of local persuasion. This can begin by making regional bodies a repository for information and policy intelligence. We suggest the following for regional bodies.

• Systematically assess the social and economic evolution of central cities and suburbs. A first step should include the establishment of a baseline, beginning in 1960 and running through the current census. Data should be collected for income, ethnic and racial disparities, levels of segregation, poverty, crime, public assistance, and levels of urban-suburban "hardship" (Nathan & Adams, 1976, 1989; Ledebur & Barnes, 1992, 1993; Savitch, Sanders, & Collins, 1992; Savitch, Collins, Sanders, & Markham, 1993; Swanson & Vogel, 1986).

• Regional governance has a better chance of success when disparities are narrower. We suggest that federal and state assistance focus on more narrowly mapped regions and build on successes over the long term. Assistance can then gradually move to the most challenging areas.

• Use existing grants and legislation to encourage regional coordination and integration. Evidence suggests that the A-95 review process and, more recently, ISTEA mandates are successful in promoting regional perspectives and facilitating intergovernmental cooperation. An incremental strategy of shifting categorical or block grant money, planning assistance, or other federal assistance to regional bodies and encouraging the formation of regional forums would have a long-term positive effect.

• Develop and disseminate a manual for local self-study and encourage the development of regional self-study commissions. The last major handbook on local government reorganization was published in 1980 by the National Academy of Public Administration. Efforts at local government reorganization continue in communities. Unfortunately, these efforts often pursue politically infeasible solutions. Local leaders lack adequate knowledge about government reorganization, its limitations, its possibilities, and its capacity to address existing problems. Public choice and metropolitan government propositions continue to misinform local debate about government structure.

- City-county consolidation is not an easy path. No new metropolitan governments have been created in medium or large metropolitan areas since 1978 (Vogel, 1994). Localities can learn a great deal from the experience of others and save a significant amount of time and effort.

- Information on local government structure, numbers, and types is difficult to track. Regional bodies can pool their resources, develop an inventory on local government structure, and promote the creation of regional commissions with a time horizon that focuses on governance in the next 25 years. Related to this, regional bodies should promote local efforts at social and economic accounting. One way of doing this is to examine composite resources available within a region and assess how they are allocated. This would include budgets from nonprofit organizations, local governments, special districts, corporate contributions, and other sources (Warren, Rosentraub, & Weschler, 1995; Long, 1986; Swanson & Vogel, 1986). By doing this, localities can better assess current and projected resources and link that assessment to policy priorities.

- Federal programs and policies can encourage regional governance or discourage it. Accordingly, we believe that there is a need to monitor the system of intergovernmental relations, so that incentives can be applied in the most effective manner. At the very least, we need to acknowledge the vast differences in state systems of local government. This would enable local actors to evaluate successes and bring about their own innovations.

- Accordingly, there should exist a federal organization responsible for coordinating, evaluating, and strengthening regionalism. These functions once were carried out by the Advisory Commission on Intergovernmental Relations (ACIR) and should be revived. Such an organization would be a valuable clearinghouse for what is now an intergovernmental system of national scope. Whether it is located in the Department of Housing and Urban Development, the ACIR, or someplace else, a first priority should be the designation of natural or political regions. These may not conform to MSA or CMSA boundaries and should result from a negotiated process with local actors.

- Finally, our case studies suggest that bi-state agencies can be effective. Indeed, many regions extend over more than one state, yet the need for congressional approval for multistate agreements constitutes an impediment to greater state cooperation. The federal government should explore successful multistate ventures and speed the political process whereby such cooperation can be encouraged.

These lessons are geared to the inherent characteristics of regionalism, as we have discerned it through 10 case studies. The lessons offer no dramatic solutions, because we do not believe that suggestions of this kind would be conducive to the nature and evolution of metropolitan regions. By its inherent characteristics, regionalism is more akin to a glacier than to a shooting star. Its movement is imperceptible, and its achievements are best framed as an evolution of progress. It is better to help guide that progress rather than to force unhappy conclusions. We would be wiser to pursue the longer, happier process.

NOTES

1. It does have zoning power in the unincorporated area of the county.

2. This was a concession necessary for the consolidation to achieve sufficient popular support.

3. New York is the only city in America that encompasses five counties, and is the country's largest metropolitan region.

Following Berg and Kantor's geographic division, we have divided the New York region into three components instead of two. These are New York County (Manhattan) as the "central county"; Bronx, Kings, Queens, Richmond, and Hudson (in New Jersey) as the "inner counties"; and the rest of the region as "suburban counties."

4. In making these calculations, we have followed each author's definition of the "functional city." Phares and Louishomme adopted a narrower definition that includes only St. Louis County and St. Louis City. Both of these encompass highly urbanized areas, and this accounts for the negative flow of commuter earnings.

REFERENCES

Advisory Commission on Intergovernmental Relations. (1993). *Metropolitan organization: Comparison of the Allegheny and St. Louis case studies.* Washington, DC: Author.

Barlow, I. M. (1991). *Metropolitan government.* New York: Routledge.

Census of governments. (1992). Washington, DC: U.S. Bureau of the Census.

Ehrenhalt, A. (1989, November). For Chambers of Commerce and cities, the days of conflict may be over. *Governing,* pp. 40-48.

Ledebur, L. C., & Barnes, W. R. (1992). *Metropolitan disparities and economic growth* (Research report). Washington, DC: National League of Cities.

Ledebur, L. C., & Barnes, W. R. (1993). *"All in it together": Cities, suburbs, and local economic regions.* Washington, DC: National League of Cities.

Long, N. (1986). Getting cities to keep books. *Journal of Urban Affairs, 8,* 1-7.

Nathan, R. P., & Adams, C. (1976). Understanding central city hardship. *Political Science Quarterly, 91*(1), 47-62.

Nathan, R. P., & Adams, C. (1989). Four perspectives on urban hardship. *Political Science Quarterly, 104*(3), 483-508.

Parks, R. B., & Oakerson, R. J. (1989). Metropolitan organization and governance: A local public economy approach. *Urban Affairs Quarterly, 25*(1), 18-29.

Peirce, N. R., Johnson, C. W., & Hall, J. S. (1993). *Citystates: How urban America can prosper in a competitive world.* Washington, DC: Seven Locks.

Rabinovitz, F. (1994, December 8-9). Comments on regional governance, Roundtables on Regionalism, Social Science Research Council and U.S. Department of Housing and Urban Development, Washington, DC.

Rothblatt, D. N., & Sancton, A. (Eds.). (1993). *Metropolitan governance: American/Canadian intergovernmental perspectives.* Berkeley: University of California, Institute of Governmental Studies Press.

Savitch, H. V., Collins, D., Sanders, D., & Markham, J. P. (1993). Ties that bind: Central cities, suburbs, and the new metropolitan region. *Economic Development Quarterly, 7*(4), 341-357.

Savitch, H. V., Sanders, D., & Collins, D. (1992). The regional city and public partnerships. In R. Berkman, J. F. Brown, B. Goldberg, & T. Mijanovich (Eds.), *In the national interest: The 1990 urban summit: With related analysis transcript, and papers.* New York: The Twentieth Century Fund Press.

Self, P. (1982). *Planning the urban region—A comparative study of policies and organizations.* University: University of Alabama Press.

Swanson, B. E., & Vogel, R. K. (1986). Rating American cities—Credit worthiness, urban distress, and the quality of life. *Journal of Urban Affairs, 8,* 139-157.

Vogel, R. K. (1992). *Urban political economy: Broward County, Florida.* Gainesville: University Press of Florida.

Vogel, R. K. (1994). *Local government reorganization.* Louisville, KY: League of Women Voters and the University of Louisville Department of Political Science.

Warren, R., Rosentraub, M. S., & Weschler, L. F. (1995). Building urban governance: An agenda for the 1990s. *Journal of Urban Affairs, 14,* 399-422.

Index

About the Authors

Bruce Berg is Associate Professor of Political Science at Fordham University. His present research interests focus on intergovernmental relations, interest groups, and health policy. He has published articles on these and related topics in *Administration and Society*, *Policy Studies Review*, and other journals.

David Brunori is the legal editor of *State Tax Notes* magazine. His research interests include state and local tax and budget issues. He earned his M.A. in political science from The George Washington University and his J.D. from the University of Pittsburgh.

Mark Ebert, an attorney, is a doctoral student in political science at The George Washington University. He has a master's degree in Public Administration and was a Banneker Fellow at the George Washington University Center of Washington Area Studies. His research interests include state and local government and political socialization.

John J. Harrigan is Professor of Political Science at Hamline University, St. Paul, Minnesota. He coauthored, with William C. Johnson, *Governing the Twin Cities Region* and is the author of *Political Change in the Metropolis, Politics and Policy in States and Communities, Politics and the American Future*, and *Empty Dreams, Empty Pockets: Class and Bias in American Politics*.

Jeffrey Henig is Professor of Political Science and Director of the Center for Washington Area Studies at The George Washington University. He is the author of *Neighborhood Mobilization: Redevelopment and Response* (1982), *Public Policy and Federalism* (1985), and *Rethinking School Choice* (1994). His articles on such topics as neighborhood organizations, gentrification, privatization, and the politics of school reform have appeared in scholarly journals including *Urban Affairs Quarterly, Journal of Urban Affairs, Political Science Quarterly*, and *World Politics*.

307

Louise Jezierski is Assistant Professor of Sociology and the Program in Urban Studies at Brown University. She was a Fellow at UCLA's Institute for American Cultures, Chicano Studies Research Center during 1992. She is writing a book, *Imagination and Consent: Reinventing the Post-Industrial City*, based on her research on the politics of the postindustrial urban transformation of Cleveland and Pittsburgh. Her other major interest is urban race and ethnic relations. She has published articles on public-private partnerships and neighborhood movements, regional development, postmodern urban theory, and the role of women in the transformation of industrial cities to service and high-technology economies. Her Ph.D. in sociology is from the University of California, Berkeley.

Paul Kantor is Professor of Political Science at Fordham University. He has written numerous articles, reviews, and books in the fields of urban politics, public policy, and political economy. Most recently, he co-authored, with Dennis R. Judd, *Enduring Tensions in Urban Politics* (1992) and wrote *The Dependent City Revisited: The Political Economy of Urban Economic and Social Policy* (1995). His current research focuses on the political economy of comparative urban development in the United States and Western Europe. He received his Ph.D. from the University of Chicago.

Claude Louishomme was formerly Director of Real Estate and Community Development for the Economic Council of St. Louis County. He also served as lead administrator for the selection of gaming developers in unincorporated areas of St. Louis County. He is a graduate student at the University of Missouri—St. Louis, pursuing a doctoral degree in political science with an emphasis on public policy, public administration, and urban politics.

Arthur C. Nelson is Professor of City Planning, Public Policy, and International Affairs at the Georgia Institute of Technology. He is widely published in the areas of regional development planning, urban and regional development patterns, resource land preservation and management, infrastructure planning and finance, growth management, and urban revitalization. He serves as an Editor of the *Journal of the American Planning Association* and Associate Editor of the *Journal of Urban Affairs*. His clients have included numerous federal agencies as well as regional, state, and local governments. He earned his doctorate in urban studies from Portland State University.

Donald Phares is Professor of Economics and Public Policy at the University of Missouri—St. Louis, and Director of the North American Institute for Comparative Urban Research, which coordinates research and conferences on cross-country urban issues. He is the author of three books and the editor of two, and he has authored or coauthored more than 100 professional articles and technical/governmental reports. He served as coeditor of *Urban Affairs Quarterly* and was on the Editorial Board of the *Journal of Urban Affairs*. He has been a consultant on government finance and urban policy issues for numerous public and private organizations, foundations, and universities.

Alan L. Saltzstein is Professor of Political Science and Coordinator of Public Administration Programs at California State University, Fullerton. He is the author of *Public Employees and Policymaking* and several articles dealing with urban politics and personnel policy making in cities, and he has also written on Los Angeles city politics. His work has appeared in the *Western Political Quarterly*, *Social Science Quarterly*, *American Review of Public Administration*, and *State and Local Government Review*. He holds a Ph.D. in political science from UCLA.

H. V. Savitch is Professor of Urban Policy and Management at the Center for Urban and Economic Research, College of Business and Public Administration, University of Louisville. He has published three books on various aspects of urban affairs, including neighborhood politics, national urban policy, and comparative urban development. His *Post Industrial Cities* (1989) was nominated for the best volume on urban politics by the American Political Science Association. He has coedited *Big Cities in Transition* (with John Thomas) and is coeditor of the *Journal of Urban Affairs*. His articles have appeared or are forthcoming in *Polity*, *Journal of the American Planning Association*, *Economic Development Quarterly*, *Urban Affairs Quarterly*, *National Civic Review*, and the *International Journal of Urban and Regional Research*.

Genie Stowers is Associate Professor of Public Administration at San Francisco State University. She has published frequently on Miami and Miami politics and is completing work on a book about Cuban American political development. Her research interests are in the areas of urban politics and policy, women and public policy, and ethnic politics. She is especially interested in the question of how politically marginalized groups develop and use power and how they work for changes in public policy.

Bert Swanson is a member of the Political Science and Urban Studies faculty and former director of the Institute of Government at the University of Florida. He served as a consultant on the charter revision of New York City and as Executive Director of the Advisory and Evaluation Committee on Decentralization for the New York City Board of Education. He has assisted public officials at the federal, state, and local levels in the areas of health, education, housing, race relations, disaster response, and civil disturbances. He is coauthor of *The Rulers and the Ruled: Political Power and Impotence in American Communities*, which won the Woodrow Wilson Foundation Award for the best book on political science published in 1964. He is the author or coauthor of numerous other books in community studies, racial and ethnic relations, and other areas. He received his M.A. and Ph.D. degrees from the University of Oregon.

Ronald K. Vogel is Associate Professor of Political Science at the University of Louisville. His research focuses on regional economic development and governance. He is the author of *Urban Political Economy: Broward County, Florida* (1992) and editor of the *Handbook of Research on Urban Politics and Policy* (forthcoming). He serves on the Executive Council of the Urban Politics section of the American Political Science Association.

DATE DUE

JUN 19 2007